Hydraulics for Operators
Revised Edition

Wm. Elgar Brown

BUTTERWORTH PUBLISHERS
Boston • London
Sydney • Wellington • Durban • Toronto

An Ann Arbor Science Book

Ann Arbor Science is an imprint of Butterworth
Publishers.

Library of Congress Cataloging in Publication Data
Main entry under title:

Water treatment plant operation.

   "An Ann Arbor science book."
   Contents:—   —[3] Hydraulics for operators /
Wm. Elgar Brown.
   Includes index.
   1. Water treatment plants—Handbooks, manuals, etc.
2. Water—Purification—Handbooks, manuals, etc.
I. Langworthy, V. W. (Virgil W.)
TD433.W3652   1985   628.3   84–29320
ISBN 0–250–40650–0 (v. 3)

Butterworth Publishers
80 Montvale Avenue
Stoneham, MA 02180

10 9 8 7 6 5 4 3 2 1

Printed in the United States of America

Contents

Series Preface

The Water Treatment Plant Operation Series includes material of interest and value to all water utility personnel. The information is presented in the context of present-day utility operation under increasingly stringent social, economic, technical, and legal considerations.

Water system employees will find the books in this series useful at all stages of the certification process. Similarly, these texts contain topical material helpful to those involved in the technical training and testing of utility personnel.

Considerable effort has been exerted to ensure that the subject matter is current and consistent with requirements throughout the United States. Where choices from among varying levels of knowledge were necessary, the most stringent option was selected.

The books in this series have been prepared from material used to train water utility personnel. We expect that the books will continue to prepare increasing numbers of people for careers in the water works industry.

Virgil W. Langworthy
Series Editor

OTHER VOLUMES IN THE SERIES

Review Manual for Operators by Wm. Elgar Brown and Richard S. Sacks.

Mathematics for Operators by E.J. Way

Chemistry for Operators by Robert L. Fountain

Preface

The flow of water and wastewater through pipes and channels, the use of pumps, and the effects of pressure exerted by water in a constructive or destructive fashion form a body of knowledge that is of vital importance to workers in the fields of water supply; treatment and distribution; and wastewater collection, treatment, and disposal.

Although the exact treatment of most problems dealing with fluids can be extremely complex, this text has as its major objective the treatment of the subject matter in as simple a fashion as possible. Therefore, the math has been simplified in this text. It is necessary that the student have the ability to operate an electronic calculator and to be able to interpolate from a nomograph.

It must be pointed out that this material can best be learned through the solution of problems. To that end, the lessons terminate with a list of problems.

The author wishes to recognize the original authorship by the Sanitary and Water Resources Engineering staff at the University of Michigan. Development of the original manuscript was supervised by Dr. Jack Borchardt at the request of the education committees of the Michigan Section, American Water Works Association and the Michigan Water Pollution Control Association.

Recognition is also accorded to Mr. Roger Zeeff, P.E., for the use of his notes resulting from teaching hydraulic courses in recent years and to Ms. Suzanne Olivier for her patience and understanding in typing the original text. The original edition had numerous errors that hopefully have been corrected in this revised edition.

Wm. Elgar Brown

Hydraulic Concepts

This chapter is designed to present the basic definitions common to the field of hydraulics as used in water supply and sewage treatment. Some common abbreviations which will be used throughout the book are:

gpm = gallons per minute
mgd = million gallons per day
cfs = cubic feet per second
psf = pounds per square foot
psi = pounds per square inch

FLUID

This text will deal with the fluid, water. A *fluid* is defined as a substance which will take the shape of its container. This is an extremely valuable property which is often the basis of metering or measuring the quantity of a flowing fluid.

Since this text deals only with water or wastewater, the term fluid shall be understood to mean water only. All gases or other forms of liquids will be excluded except when used to illustrate a difference. At such times, the materials will be defined by name.

In this text a fluid must be considered incompressible. Thus, a cubic foot (0.03 m^3) of water will occupy a volume of one cubic foot regardless of how much pressure is applied to it.

WEIGHT

Weight is defined as the force of the earth's gravitational pull upon a body. A falling body increases its speed as it falls to the earth. As a result, there is an increase in speed at the rate of 32.2 feet per second per second (ft/sec^2) or 9.81 m/sec^2. This term is referred to as *g* and means that a falling body will increase in speed 32.2 feet per second every second that it falls.

We are frequently concerned with the weight of a substance, or a definite quantity of fluid. The attraction of the earth for the substance will cause the material to have weight, and this weight will be considered to be the same wherever this material is placed, regardless of any change in its physical shape. The unit in which weight is most frequently expressed is the pound (lb), or kilogram (kg) (2.2 lb). For large weights, the ton (2000 pounds) is used.

A man may know that he weighs 160 lb (72.7 kg), but he would never know what volume he occupies. He buys meat by the pound, but he buys milk by the quart or gallon and his water is paid for on the basis of gallons or cubic feet.

Thus, volume is also an important aspect of a fluid. The concept of the cubic foot is merely a box one foot on each edge. In the study of hydraulics, the student jumps back and forth between volume of water and weight of water, and he must know how to make the conversion.

The weight of a substance per unit volume is defined as its *specific weight*. One cubic foot of water weighs 62.4 lb (28.4 kg). Thus, the specific weight of water is 62.4 lb/ft^3. If our 160-lb man were considered to be all water contained within a thin weightless membrane, then his volume would be 160/62.4 = 2.56 ft^3.

Example 1–1

If the specific weight of a given sand is 105 lb/ft^3, how much does a cubic yard (yd^3) of sand weigh?

$$1 \text{ yd}^3 = 3 \text{ ft} \times 3 \text{ ft} \times 3 \text{ ft} = 27 \text{ ft}^3$$

$$27 \text{ ft}^3 \text{ weighs } 27 \times 105 = 2835 \text{ lb/yd}^3$$

How many cubic feet of the above sand would be found in one ton of sand, and how many cubic yards would this be?

$$\text{One ton} = \frac{2000}{105} = 19 \text{ ft}^3$$

$$\text{One ton} = \frac{19}{27} = 0.704 \text{ yd}^3$$

Most fluids other than water can be compared to pure water with regard to their weight per unit volume. This comparison is known as *specific gravity* (SG), and is a ratio of the weight per unit volume of the other fluid compared to water. Mercury, sometimes used in water works instrumentation, weighs 849 lb/ft³.

$$\text{SG of mercury} = \frac{849.0}{62.4} = 13.6$$

ELEVATION

The term *elevation* is often encountered in water works practice. It usually means the height of a point, this height being measured in feet above a datum plane. The usual datum is mean sea level. Thus, an elevation of 860 means 860 feet above sea level. In the case where no known reference to sea level is available, elevation is referred to some other easily obtained datum, such as the main floor of the water plant. This datum is given an arbitrary elevation, such as 700 (meaning, let us assume this floor is 700 ft above sea level). Then all points can be set above or below 700 by the correct distance.

It must be noted that no matter whether the datum is sea level or merely an arbitrarily assumed value, the actual vertical distance in feet between two points is given by subtracting the lower elevation from the higher elevation.

The term *head* is also used in the field of hydraulics. It usually refers to a height in feet of water above the point in question; for example, "an opening in the side of a tank is discharging water under a head of ten feet." The head would be measured from the centerline of the opening to the water surface above. If one knew the elevation of the water surface and the centerline of the opening, then the head would be the difference in the two elevations.

Later in the text the term *head* will be shown to involve energy. The aspects of motion and the loss of energy in friction due to motion change the simple picture of head as described above. For the present, it is sufficient to know that *head* means feet of water above a point. This term in many cases is synonymous with pressure. (Pressure will be discussed in Chapter 2.)

AREAS

Areas of a square or rectangle are calculated by multiplying the length by the width and expressing the product in square feet. An area in square feet may be converted into acres (ac) by dividing by 43,500 ft²/ac.

In calculating the area of a circle, the terms radius and diameter are needed. The diameter of a circle is twice the radius. The radius is the distance from the center of the circle to the outside edge of the circle. The area (A) of a circle is equal to pi (π), which is 3.14, times r^2, or $A = \pi r^2$. Since $r = d/2$ we can substitute for r^2 the term $(d/2)^2$ and the equation becomes 3.14 $\times (d/2)^2$, or the area of a circle is equal to 3.14 $(d^2/4)$.

The distance around the edge of a circle is known as its circumference. It can be computed using the equation for the circumference of a circle, $c = \pi d$. It is possible to find the diameter of a pipe when the pipe is closed by measuring the distance around the outside of the pipe and, with allowance for the pipe wall thickness, determine the inside diameter.

The above information is most often used in finding the cross-sectional area of circular pipes. If one has a free length of pipe, the easiest measurement is a diameter, and the area is simply $A = (3.14/4)d^2$. The area is usually desired in square feet, so d must be converted to feet.

Example 1–2

A pipe has an inside diameter of 2 in. What is its area in square feet?

First, convert inches to feet by:

$$d = \frac{2}{12} = .17 \text{ ft}$$

$$A = \frac{3.14}{4} (.17)^2 = 0.022 \text{ ft}^2$$

Frequently situations exist where diameter cannot be measured.

Example 1-3

What is the area of a pipe which is erected and cannot be taken down? In this case, one can measure the circumference and compute the outside diameter, allowing for pipe thickness; select a logical inside diameter; and then compute the area.

A cast iron gallery pipe measures 54.5 in in circumference. What is its inside area?

$$c = 3.14 \text{ d; so d} = \frac{54.5}{3.14} = 17.4 \text{ in}$$

Diameter (out to out) = 17.4 in (allowing for wall thickness and knowing pipe diameters are in even inches, inside diameter must be 16 in).

$$d = \frac{16}{12} = 1.33$$

$$A = \frac{3.14}{4} (1.33)^2 = 1.39 \text{ ft}^2$$

VOLUME

The volume (V) of a rectangular tank would be most easily computed by measuring its dimensions and determining the product of L × W × D = Vol. Thus, a tank 100 ft long, 10 ft deep, and 40 ft wide would contain:

$$V = 100 \times 10 \times 40 = 40,000 \text{ ft}^3$$

Likewise, such a tank would contain: (1 ft³ = 7.48 gal)

$$V = 40,000 \times 7.48 = 299,200 \text{ gal}$$

Example 1–4

If water is flowing into the above tank at the rate of 100 gpm, how long will it take to fill the tank?

$$\frac{299,200}{100} = 2992 \text{ min}$$

$$\frac{2992}{60} = 49.8 \text{ hr}$$

$$\frac{49.8}{24} = 2.08 \text{ days}$$

The volume of a cylindrical tank is determined by calculating the product of the area of the base and the height of the tank. If one remembers to use A = (3.14/4)d² (where d is measured in ft), then the volume = (3.14/4)d²h.

Example 1–5

A standpipe has a diameter of 30 ft and a height of 80 ft. How many ft³ does it contain and how many gallons?

$$V = \frac{3.14}{4} (30)^2 (80)$$

$$V = 56,500 \text{ ft}^3$$

$$V = 56,500 \times 7.48 = 423,000 \text{ gal}$$

It should be noted that a pipe is, in effect, a long cylinder.

Example 1–6

How much water can be stored in a 16-in pipe that is 120 ft long?

$$d = \frac{16}{12} = 1.33 \text{ ft}$$

$$V = \frac{3.14}{4}(1.33)^2(120) = 167 \text{ ft}^3$$

$$167 \text{ ft}^3 \times 7.48 \text{ gal/ft}^3 = 1250 \text{ gal}$$

In the measurement of flow at a gauging station along a river, the unit second-feet is often used. For example, the average discharge of a river during a one-month period was 38 second-feet. This terminology is synonymous with cfs and should always be so interpreted.

A somewhat different unit of volume is the ac-ft, which is one acre of area covered with one foot of water. This term is applied to such items as earth reservoir storage. For example, a reservoir has a capacity of 1500 ac-ft. Since one acre equals 43,560 ft^2, the ac-ft must be equal to 43,560 ft^3, or 43,560 $\times$ 7.48 = 326,000 gallons.

The reservoir above therefore has a storage of:

$$1500 \times 43,560 = 65,340,000 \text{ ft}^3$$

$$1500 \times 43,560 \times 7.48 = 488,743,200 \text{ gal}$$

Now from the size of the numbers it can be seen that the ac-ft is, in reality, a device to record fairly large numbers in a recognizable but simpler form.

The term *cubic feet* (ft^3) has previously been defined. A subdivision of this unit is the gallon. Though gallons are based on a definite number of in^3, it is best to deal only with the relationship of gallons to cubic feet (7.48 gallons = 1 cubic foot); since 1 cubic foot weighs 62.4 lb, a gallon must weigh 62.4/7.48 = 8.34 lb.

UNITS OF FLOW

The *units of flow* are the means by which a volume of water passing a particular point in a unit of time is measured. The most frequently used measures are the cubic foot per second (cfs), which is one cubic foot of water passing a point every second, and gallon per minute (gpm), which is one gallon of water passing a point every minute. One cubic foot per second is equal to 448.8 gallons per minute (1 ft^3/sec × 60 sec/min × 7.48 gal/min = 448.8 gal/min). The unit million gallons per day (mgd) is also used. This is the flow in million gallons of water flowing past a point each day.

Most pumps are sold with a discharge rating in terms of gpm. Sand filters are sometimes rated in terms of million gallons per day. Water rates are quoted on the basis of cost per thousand gallons or hundred cubic feet.

Example 1–7

A 100-gpm pump will pump how many cubic feet per second (cfs)?

$$Q = \frac{100}{7.48} = 13.4 \text{ ft}^3/\text{min}$$

$$= \frac{13.4}{60} = 0.223 \text{ cfs}$$

Flow rate (Q) will be discussed in more detail in Chapter 3.

SLOPE

Since it is not always possible or desirable to lay pipe in a level or horizontal position, we use the word *slope* to describe the rise or fall of the pipe with respect to horizontal distance. This vertical rise or fall is measured in feet. If the elevation of each end is known, the rise or fall can be found by subtracting one elevation from the other. Thus, a pipe, one end of which is one foot higher than the other, is said to slope. One way of expressing the slope is as a percent. If the pipe mentioned above is 20 ft long, the rise or fall (1 ft) occurs in a horizontal distance of approximately 20 ft (the length of the pipe) and its slope is 1/20, or 5%.

HYDRAULIC RADIUS

When water flows in a pipe, the water rubs against the sides of the pipe, resulting in what is known as *pipe friction*. This friction, and the head loss which results from it, is described more fully in Chapter 5. In the calculations of the flow through pipes or other conduits, we find a term known as the *hydraulic radius*. The hydraulic radius is a measure of the efficiency with which a conduit can transmit water. By definition, the hydraulic radius is the wetted area of flow divided by the wetted perimeter. (The wetted perimeter is the length of the cross section of the conduit which is in contact with the water.)

For example, the hydraulic radius of a 12-in-diameter pipe flowing full is found as follows:

Wetted area of flow = cross sectional area of pipe

$$= \frac{\pi(d)^2}{4} = \frac{3.14 \times 1}{4} = 0.785 \text{ ft}^2$$

Wetted perimeter = circumference of the pipe

$$= \pi d = 3.14 \times 1 = 3.14 \text{ ft}$$

$$\text{Hydraulic radius} = \frac{\text{Wetted area}}{\text{Wetted perimeter}}$$

$$= \frac{\pi(d)^2/4}{\pi d} = \frac{.785}{3.14} = 0.25 \text{ ft}$$

As can be seen by the previous example, the hydraulic radius for a circular pipe flowing full is d/4.

For the best efficiency in transmission of fluids it is necessary to have the greatest area of flow with the least amount of contact with the wetted surface of the conduit (wetted perimeter).

A circular pipe gives the maximum flow area for the least wetted perimeter for a totally enclosed conduit. As an illustration, let us assume an area of flow of 2 ft² and determine the hydraulic radius of various shapes with this cross sectional area, all flowing full.

Example 1–8

Circular pipe:

$$A = \frac{\pi(d)^2}{4} = 2 \text{ ft}^2$$

Solving for d gives, d = 1.595 ft. (Since this is an arbitrary area, the diameter is not necessarily an available pipe size.)

$$\text{Hydraulic radius} = \frac{d}{4} = \frac{1.595}{4} = 0.399$$

Square channel:

$$\text{Area of flow} = 2 \text{ ft}^2 = h \times w = 2 \text{ ft}^2 \text{ (closed at top)}$$

$$h = w \text{ (for square channel)}$$

$$h^2 = 2$$

$$h = 1.414 = w$$

Wetted perimeter $= 2h + 2w$
$$= (2 \times 1.414) + (2 \times 1.414) = 5.656$$

$$\text{Hydraulic radius} = \frac{2}{5.656} = 0.354$$

Rectangular channel where $w = 2h$:

$$A = 2 \text{ ft}^2 \text{ (closed at top)}$$

$$A = wh = 2 \quad \text{or} \quad 2h^2 = 2; \quad h = 1 \text{ ft} \quad \text{and} \quad w = 2 \text{ ft}$$

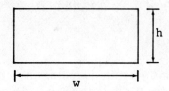

$$\text{Wetted perimeter} = 2w + 2h = 6$$

$$\text{Hydraulic radius} = \frac{2}{6} = 0.333$$

The same type of relationships can be shown comparing flow in half of a circular pipe with flow in square or rectangular channels which are open at the top. As shown by these calculations, the circular shape produces the highest hydraulic radius. The larger the hydraulic radius the more efficiently (with lower head loss) the water is carried by the conduit.

PROBLEMS

1–1. If a tank 150 ft long, 10 ft deep, and 35 ft wide is full of water, how many gallons does it contain? A pump which is *rated* at 100 gpm is being used to pump out this tank, and it takes exactly three days to empty it. What is the average pumping rate?

Answer: 392,700 gal, 90.9 gpm

1-2.   Determine the hydraulic radius for each of the following:

    a.   24-in i.d. circular pipe flowing full.
    b.   24-in i.d. circular pipe flowing half full.
    c.   Rectangular section, 3 ft × 2 ft closed at the top, all surfaces wet.

    Answer: a.   0.50 ft
                  b.   0.50 ft
                  c.   0.60 ft

1-3.   If the elevation of the bottom of one end of a channel is 732.48 ft and the other end has an elevation of 730.22 ft and the channel is 57.3 ft long, what is the slope of the channel?

    Answer: 3.94%

1-4.   A circular tank 15 ft in diameter has a flow into it at the rate of 75 gpm. If the tank is 10 ft deep, how long will it take to fill if it has 3 ft of water in it at the beginning?

    Answer: 123 min

1-5.   Compute the specific weight of each of the following, in lb/ft^3

    a.   3 yd^3 of material weighing 1475 lb
    b.   5 gal of material weighing 75 lb

    Answer: a.   18.2 lb/ft^3
                  b.   112.2 lb/ft^3

1-6.   Compute the specific gravity of each of the following:

    a.   Alcohol, which weighs 6.65 lb/gal
    b.   Gasoline, which weighs 7.5 lb/gal

    Answer: a.   0.797
                  b.   0.899

1-7.   A pail of water weighs 40.0 lb and the empty pail weighs 2.5 lb. How much water (in gallons) is in the pail?

    Answer: 4.5 gal

1–8.   It is necessary to determine the volume of water contained in a section of a pipe line 8 in in diameter and 1000 ft long for disinfection. What is the volume?

Answer: 349 ft³, or 2610 gal

2

Pressure

Pressure is a basic hydraulic concept that water and sewage works personnel frequently encounter. Fluids, such as water and air, have weight and therefore exert pressure or force upon surfaces of contact. In another sense, pressure can be thought of as the tendency for a fluid to force its way out of confinement. Pressure can be presented in two ways: first, as *total pressure* which relates to the total force exerted by a fluid on the surface under consideration; second, and more commonly, as *unit pressure* which is a measure of the force exerted by a fluid upon a unit area of the surface under consideration. This unit pressure may also be thought of as a pressure intensity.

TOTAL PRESSURE

The student should begin by developing the concept of total pressure being equivalent to weight, but only when there is little or no motion of fluid. As an example, assume a cubical steel tank, 10 ft on each edge, that is filled with water. From Chapter 1, the specific weight of water is known to be 62.4 lb/ft^3.

$$\text{Volume of tank} = 10 \text{ ft} \times 10 \text{ ft} \times 10 \text{ ft} = 1000 \text{ ft}^3$$

$$\text{Total weight of water} = 1000 \times 62.4 = 62,400 \text{ lb}$$

The total weight of water must be supported by the bottom of the tank. This would be 62,400 lb, which must also be the total pressure of the water on the tank bottom.

This fact would be useful in design of the tank or of a structure designed to hold the tank. It is not used as often as unit pressure which has application in pipe flow, pumps, and more complex pressure computations.

UNIT PRESSURE

Since the total pressure exerted by the water in the previous example is pressing down on a bottom area of tank which is 10 ft wide and 10 ft long, or 100 ft² in area, it is clear that the pressure on any one ft of area is:

$$\frac{62,400}{100} = 624 \text{ lb}$$

The tank may be thought of as containing 100 columns of water 1 foot square and 10 feet high, each weighing 624 pounds. Unit pressure is really total pressure exerted on a unit area, such as pounds per square foot. Expressing this in the form of an equation:

$$p = wh$$

where h = depth of water above the point, ft
      w = weight of a ft³ of water, lb/ft³
      p = unit pressure, lb/ft²

In the previous example, in answer to the question, "what is the unit pressure at the bottom of the tank?", the most direct solution would be:

$$p = wh$$

$$p = 62.4 \text{ lb/ft}^3 \times 10 \text{ ft}$$

$$p = 624 \text{ lb/ft}^2$$

It must be noted that units are extremely important here. In this example, the entire problem was carried out in terms of feet.

Unit pressures are more often used as pounds per square inch (psi). If each square foot of the tank bottom is subdivided into (12 × 12) 144 columns of water 1 sq in in cross section and 10 ft high, then the 624 pounds on each square foot would have to be divided by 144 to find the weight on each square inch.

$$\frac{624}{144} = 4.33 \text{ psi}$$

From this it can be noted that a problem which gives values in psi must have these values converted to pounds per square foot by multiplying by 144 before applying the formula p = wh.

Example 2-1

A gauge on the side of a tank indicated a pressure of 38.4 psi. How high was the water level inside the tank? The pressure adjacent to the gauge was:

$$p = 38.4 \times 144 = 5530 \text{ lb/ft}^2$$

From p = wh

$$h = \frac{p}{w} = \frac{5530}{62.4} = 89 \text{ ft above the gauge}$$

The utility of the above equation is thus demonstrated where pressure problems are involved, but units must be carefully observed. Actually, many gauges are calibrated directly in feet of water and in psi at the same time. The direct conversion from psi to feet of water head will be demonstrated later in this chapter.

Example 2-2

A well must be studied for drawdown characteristics. Copper tubing is inserted into the casing to the well bottom and fitted at the surface with a gauge and valve connection for introducing compressed air. When the submersible pump is turned off, the maximum reading which can be developed by applying air pressure is 52 psi. When the submersible pump is discharging 500 gpm, the maximum air pressure which can be developed is 29.5 psi. What is the drawdown h at that time?

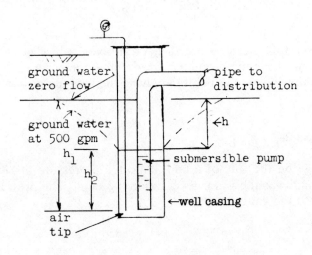

$$p_1 \text{ at zero flow} = 52 \times 144 \text{ lb/ft}^2$$

$$p_2 \text{ at 500 gpm} = 29.5 \times 144 \text{ lb/ft}^2$$

$$h = h_1 - h_2$$

Since $p = wh$ and $h_1 = p_1/w$, while $h_2 = p_2/w$, then:

$$h = \frac{p_1}{w} - \frac{p_2}{w} = \frac{52 \times 144}{62.4} - \frac{29.5 \times 144}{62.4}$$

$$h = 52 \text{ ft}$$

Remember p_1 and p_2 are the pressures at the bottom of the air line.

PASCAL'S PRINCIPLE

Reflection on the first problem above shows a slightly different twist to the idea of weight causing pressure. A gauge connection was drilled into the side of a tank. Since the hole was at right angles to the water column adjacent to it, certainly the water did not rest on the hole. Pascal's principle states that at a point at rest in a fluid the pressures in all directions are equal.

Therefore, if a glass tube had been inserted into the gauge hole and bent up along the side of the tank, the water pressure at the hole would have caused the water to rise in the tube until the water level in the tube was exactly equal to that within the tank.

Such tubes or sight gauges are frequently used where tank height is small. In Example 2–1 a gauge was used because an 89-ft glass tube would be too hard to install and difficult to read.

Pascal's principle has other applications, too. If the pressure is the same in all directions, then it must act upward as well as downward.

Example 2–3

A pipe below a basement floor breaks, releasing water. The water flows upward around the basement walls and away at the ground surface 12 ft above. What unit pressure is acting to buckle the basement floor?

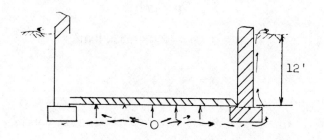

$$p = wh = 62.4 \times 12 = 748.8 \text{ lb/ft}^2$$

This same kind of pressure acts to float tanks submerged in groundwater and is the reason for the pressure relief valves frequently seen in such tanks.

CONVERSION OF UNIT
TO TOTAL PRESSURE

The application of the unit pressure equation often comes through the conversion of such values into total pressure. The direct approach is merely to state that total pressure or force (F) is unit pressure multiplied by the area over which it acts, or in equation form:

$$F = pA$$

where F = total pressure or force
p = unit pressure
A = area in appropriate units

Example 2–4

A standpipe has a hole for a gauge connection, with an area of 0.6 in². What force would act to blow a cork out of that hole if the gauge showed 89 ft of water in the tank?

$$p = wh$$

$$p = 62.4 \times 89 = 5550 \text{ lb/ft}^2$$

$$p = \frac{5550}{144} = 38.5 \text{ psi}$$

$$F = pA = 38.5 \times 0.6 = 23.1 \text{ lb on the cork}$$

It seems obvious that if one has a gauge reading in feet or psi, the problem in this example could be solved in the easiest fashion if the 38.5 psi reading were used rather than the 89 feet. In other words, if total pressure or force on an area given in square inches is desired, a reading of unit pressure in psi gives the easiest answer. Where total pressure or force is required on an area given in square feet, then probably the pressure in lb/ft² is the easiest unit to use.

The pressure or force acting to blow a cap or blind flange off a pipe would be computed using the same techniques as illustrated for the cork above. First, the average unit pressure would be obtained at the pipe center expressed in pound per square foot from p = wh and then the product F = pA is obtained. The student must be sure that the maximum area over which this pressure can act is utilized. For example, in the case of the blind flange, the maximum area is determined by d_2, the inside diameter of the gasket (see sketch) and not d_1, the pipe diameter.

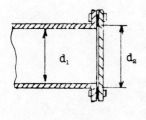

Blind Flange

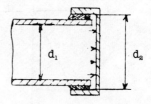

Pipe Cap

In the case of the pipe cap, the pressure again is a product of p and A, where A again is a function of d_2 which may be much larger than d_1, the inside diameter of the pipe.

VARIABLE PRESSURES

Pressure is constant over the area in question only when the head to all portions of the area is the same. This would be true on the bottom of a tank where $h_1 = h_2 = h_3$.

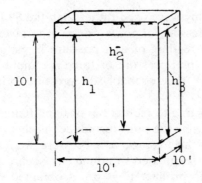

$h_1 = 10$ ft

liquid is water

This would not be true if the question asked were, "what force is trying to push out the side of the box illustrated?" Looking at a cross section of one side:

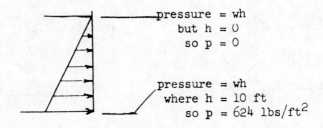

pressure = wh
but h = 0
so p = 0

pressure = wh
where h = 10 ft
so p = 624 lbs/ft^2

The average unit pressure on the tank side is $(0 + 624)/2 = 312$ lb/ft^2. Using the average unit pressure, the total force is:

$$F = Ap_{avg}$$

$$F = 10 \times 10 \times 312 = 31,200 \text{ lb}$$

An average force can be computed for any area which is square, circular, or rectangular, without regard to the angle such surface makes with the water surface. Such a force must be thought of as acting normal (or perpendicular) to the surface area in question.

SURFACES OTHER THAN SQUARE, CIRCULAR, OR RECTANGULAR

A nonsymmetric surface must be handled by complex mathematics unless it is horizontal so that h is constant for all points on the surface, or unless the pressure head is great in comparison to the area of the surface.

PRESSURE AT THE SAME DEPTH IN A CONTINUOUS FLUID IS THE SAME

Pressure is usually desired at the center of a pipe or at a point of connection to a tank. However, it must be stressed that a gauge reads pressure at the center of the gauge. If a gauge must be placed above the pipe some distance for some reason, such as convenience in reading, then a correction must be made to the gauge reading so as to reveal the correct value at the centerline of the pipe or connection to the tank.

For example, assume a pipe lies below a grating in a dry well. A gauge connection must be carried under a pipe and up the wall to be read easily. What is the true pressure in the pipe? First, a careful job of measuring from center of pipe to center of gauge is done. Let this be 10 ft. Suppose p = 10 psi at the gauge.

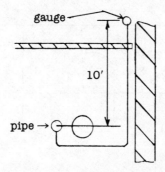

The pressure at the gauge center is 10 psi. This is equivalent to 23.1 ft of water, and a hole drilled into the pipe with a glass tube inserted would show water rising 23.1 ft above the gauge center. This would be 33.1 ft above the pipe center since the pipe is 10 ft below the gauge.

The rule is that gauges placed above the connection point have a correction in feet added to the gauge reading. Gauges placed below the connection point have the correction subtracted from the gauge reading. The magnitude of the correction is a value in feet measured vertically from the point in question to the gauge center. Obviously, such connections can be full of bends and these will not affect the result as long as a bleed off point is provided at the gauge to make sure no air is trapped in the gauge connection.

By this time, the reader must realize that water pressure can be expressed in feet, psi, or pounds per ft². All these values mean the same thing and can be converted from one to another. Imagine one cubic foot of water. This imaginary cube would be one foot high and would be resting on one square foot or 144 square inches. Since one cubic foot of water weighs 62.4 lb, the pressure from the one foot of head would be 62.4 lb/ft² or 62.4/144 = 0.433 lb/in². Therefore, to convert feet of water head to psi, multiply the feet of water by 0.433. Conversely, to convert psi to feet of head, multiply the psi by 2.31 (1/0.433 = 2.31).

Example 2–5

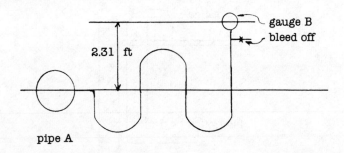

Pipe A has a pressure of 5 psig at its center. What will gauge B read?

$$2.31 \text{ ft} = 1 \text{ psi}$$

$$pg + corr = p \text{ (at pipe)}$$

$$pg + 1.0 = 5$$

$$pg = 4 \text{ psi}$$

ATMOSPHERIC PRESSURE

The atmosphere is composed of gases, which have weight and, therefore, exert pressure just as was illustrated above in the case of water. In fact, man walks around on the earth's surface submerged in a sea of air about 200 miles in depth. Atmospheric pressures are usually referenced to a standard zero datum, which is sea level. Under this zero datum condition at sea level, the atmosphere exerts an approximate pressure of 2116 lb/ft^2 or about 14.7 psi. Standard atmospheric pressure happens to be equalized by a column of mercury 30 in high since the specific gravity of mercury (Hg) is 13.6. This same standard pressure is also equalized by 34 ft of water.

Atmospheric pressure readings that are reported by local weather stations are usually expressed in inches of mercury. An increase in elevation above sea level decreases the local atmospheric pressure, since the height of atmospheric air above the ground surface is decreased. Table 2–1 illustrates the variation in atmospheric pressure with increasing elevation above sea level.

ABSOLUTE PRESSURE AND GAUGE PRESSURE

Pressure readings can be reported either as absolute pressure or as gauge pressure. The common notation for these pressures is the addition of an "a" after the pressure units for absolute pressure (i.e., psia) and the addition of a "g" for gauge pressure (i.e., psig). *Absolute pressure* refers to pressures above a datum of absolute zero, and is always a positive value. *Gauge pressure* is the pressure read directly from pressure gauges. Values can be positive or negative, or, in other words, above or below atmospheric

Table 2–1 Variation of Absolute Pressure with Altitude

Altitude (ft)	Absolute Pressure		Feet of Water
	(in Hg)	psi	
0	30.0	14.7	34.0
200	29.8	14.6	33.8
400	29.6	14.5	33.5
600	29.4	14.4	33.3
800	29.2	14.3	33.0
1,000	28.9	14.2	32.8
1,500	28.3	14.0	32.4
2,000	27.8	13.7	31.7
2,500	27.3	13.5	31.2
3,000	26.8	13.2	30.5
4,000	25.8	12.7	29.4
5,000	24.8	12.2	28.2
6,000	22.5	11.8	27.3
8,000	21.2	11.0	25.4
10,000	20.9	10.3	23.8

pressure. Under these pressure considerations then, standard atmospheric pressure at sea level is equal to 14.7 psia and 0 psig.

The value of gauge readings is obvious. The effect of atmospheric pressure is canceled out since, whatever the atmospheric pressure may be, the gauge reading is merely a value above or below the datum established by the existing atmospheric pressure.

The student must be cautioned about negative pressure values. Hydraulic computations might produce a negative value of any magnitude, but since the atmospheric pressure is a maximum of 34 feet of water at sea level, negative values greater than this are impossible. The vapor pressure of water produces an actual value somewhat less than 34 feet, but for all practical purposes, vapor pressure can be neglected in most hydraulic computations.

In this text, then, further references to pressure will be in terms of gauge pressure in psi or feet of water. Such a procedure eliminates the variable of atmospheric pressure which can change with weather or altitude.

A legitimate negative gauge pressure means that an actual vacuum of some magnitude exists at that point. Obviously, if a fluid such as air or water could rush in and adjust the pressure in that space to the existing atmospheric pressure, it would do so.

Negative pressures are frequently developed in suction lines of pumps, at the throat of diverging sections in pipe lines, and at the summit of siphons. The student merely handles such a situation by use of a negative sign which indicates that a vacuum measured below the atmospheric datum exists at the point in question.

PROBLEMS

2–1. A vertical gate in the face of a dam is 4 ft wide and 6 ft high and hinged at its upper edge. The gate is kept closed by the pressure of water standing 8 ft deep over its top edge. What total force acts to keep the gate closed?

Answer: 16,473.6 lb

2–2. If a 12-inch flanged pipe is closed at the end by a blind flange, what total stress will be carried by the bolts when the head on the pipe center is 240 ft? Assume the gasket has a 12-inch inside diameter.

Answer: 11,756.2 lb

2–3. A rectangular caisson is to be sunk to form the foundation for a bridge pier. It is in the form of an open-top box, 50 ft × 20 ft and 23 ft deep. If it weighs 75 tons, how deep will it sink when launched? If the water is 20 ft deep, what *additional* load will sink it to the bottom?

Answer: 2.4 ft, 549 tons

2–4. A steel drum which weighed 60 lb was placed on a scale and was nearly filled with water. A total load on the scales of 320.25 lb was read. How many pounds of water were actually in the drum? (It was discovered that a 3-inch steel shaft suspended from the ceiling above had its lower end immersed in the water to a depth of 1 foot at the time of weighing.)

Answer: 257.19 lb

2–5.   A piece of yellow pine lumber weighs 40 lb per ft³ and measures 6 in × 12 in × 20 ft. What maximum load will it carry if just brought to the verge of sinking in freshwater?

Answer: 224 lb

3

Fundamentals of Flow I

Several different types of flow may occur in pipes and other conduits; the type of flow depends mainly on the speed or velocity at which the water moves and on the shape and roughness of the conduit. Therefore, it is common practice to classify flow according to the type of conduit and according to the speed of flow.

If the conduit is closed except at its ends and is completely filled with the flowing water, the flow is called *pipe flow* (for example, flow in a water main or a fire hose). If the conduit is not closed, or is closed but not filled by the flowing water, the flow is called *open flow* or *open-channel flow* (for example, flow in an open wastewater trough, flow through a settling basin, and flow such as normally occurs in sewers).

Further classification is made according to two ranges of speed or velocity of water flow, either of which may occur in both open-channel flow and pipe flow; namely, *laminar* flow and *turbulent* flow. When the velocity or speed of motion of the water is very low, the individual water particles move along straight paths or lines of flow in a conduit, and the flow is said to be laminar. For laminar flow in straight conduits, the flow lines are all straight and parallel to each other and to the walls of the conduit. In curved conduits and when passing obstacles in the path of the flow, the flow lines also remain orderly and do not cross each other. An example of the behavior of flow lines for laminar flow is given in Figure 3–1. When the speed of motion becomes great enough, the individual water particles follow winding, irregular paths, and the flow is termed *turbulent*. An example of the be-

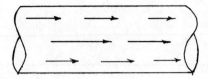

Figure 3–1.

havior of lines of flow under turbulent conditions is shown in Figure 3–2.

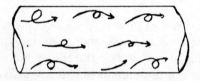

Figure 3–2.

Laminar flow occurs frequently in the passage of water through soils, sands, gravels, or porous solids such as in sand filtration. However, most flows in pipes and other hydraulic structures used in the operations of water and wastewater collection, treatment, and distribution or disposal are generally turbulent.

CONTINUITY EQUATION

For a fixed rate of flow (constant volume of water per unit of time) in a conduit, the velocity of flow (distance traveled per unit time) will depend on the size of the conduit. For a given flow in a pipeline which is made up of sections of different diameter, the velocity will be greater in the sections of small diameter than in the sections of large diameter. This can be stated in equation form as:

Flow (Q) = Cross Sectional Area (A) × Velocity (V)

$$Q = AV \qquad\qquad (3.1)$$

Equation 3.1 is called the *equation of continuity* because it is a statement of the fact that the quantity in cubic feet per second is equal to the products of the velocity in feet per second and the cross sectional area of liquid flowing in square feet. If no water enters or leaves the pipe(s), the quantity of water flowing from point to point is the same, and Q at point 1 must equal Q at point 2.

$$Q_1 = Q_2$$

Since $Q_1 = A_1 V_1$ and $Q_2 = A_2 V_2$, $A_1 V_1 = A_2 V_2$. This equation is used whenever there are varying sizes of conduit.

An illustration of a pipe with different diameters (therefore different cross sectional areas) is given in Figure 3–3. The flow, Q_1, into the pipe must be the same as the flow, Q_2, out of the pipe.

Therefore, if the cross sectional area (A_1) of the smaller of the two pipes in Figure 3–3 is one square foot and that of the larger pipe (A_2) is two square feet, and if the flow, Q_1, is equal to 2 cfs, then the velocity (V_1) in the smaller section is $V_1 = Q_1/A_1 = 2$ cfs/ft^2 = 2 ft/sec and the velocity (V_2) in the large section is $V_2 = Q_2/A_2 = 2$ cfs/2 ft^2 = 1 ft/sec. Thus, the smaller the diameter of the pipe, the greater the velocity must be to maintain a constant flow.

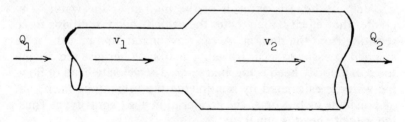

Figure 3–3.

HEAD

Water flows because there is energy or a driving force to make it flow. This driving force is called the *head* on the water. There

are three forms of head: *pressure, elevation,* and *velocity head.*
All three forms of head or energy must be considered each time
a problem in flow is considered. For convenience in handling the
mathematical computations, all forms of head are expressed in
terms of feet of water. (Pressure head has been discussed in Chapter 2.)

In a problem involving pressure head in pipe flow, pressure
head is always synonymous with the height to which water would
rise in a glass tube inserted in the pipe at right angles to the
flowing water, and is equal to p/w = h where p is unit pressure
in lb/ft² and w is the weight of the water in lb/ft³.

Elevation head is the driving force or energy resulting from
the elevation of the water with respect to some reference level or
datum. The higher the elevation, the greater the elevation head
will be. For example, a brick dropped from a third floor window
of a building will hit the sidewalk (which is the datum in this
example) with a greater force than will the same brick dropped
from the first floor window. If the brick is lying on the sidewalk,
its elevation head is zero. If the sill of the third floor window is
25 ft above the sidewalk, a brick lying on the sill has an elevation
head of 25 ft. It is common practice to measure all heights from
a datum or reference level below the lowest point of a pipe or
other hydraulic structure. Thus, all vertical distances can be measured upward from the datum, avoiding the use of plus and minus
signs to indicate the direction of measurement.

Velocity head is a result of the motion of the water. The
greater the velocity, the greater the velocity head or driving force
resulting from the motion. A car traveling at 50 mph cannot be
stopped as easily as a car traveling at 10 mph because the driving
force of velocity head is much greater. The velocity head of flowing water is calculated by multiplying the velocity by itself, V^2,
and dividing by two times the acceleration due to gravity, g. Thus,
the velocity head is equal to:

$$\frac{V^2 \ (ft/sec)^2}{2g \ (ft/sec^2)} = \frac{V^2}{64.4} \ (ft)$$

The total head or energy causing flow in a hydraulic system
is the sum of the pressure, elevation, and velocity head. Each of

these quantities may vary up or down from point to point in a system, but the sum of the three sources of energy must always be related.

If flow is taking place from one point to another, there must be an expenditure of energy to create that flow. This means that a decrease in energy must take place along the line of flow of a fluid which is exactly equal to the work done in moving the fluid from point to point. The more fluid is moved, the more the energy must be expended (assuming the same efficiency in moving the fluid prevails).

Conversely, if the sum of energy from point to point indicates an equality, obviously the flow is zero.

If a computation indicates an increase in energy from point one to point two, the student may be sure that a mistake had been made in his assumptions or computations or that a pump lies between points one and two which is inserting energy into the system.

Example 3–1

A flow, Q, of 2000 gpm takes place in a pipe with an inside diameter, d, of 12 in. What is the velocity, V, of flow in the pipe?

Solution:

From Equation 3.1, Q = AV

$$1 \text{ ft}^3 = 7.48 \text{ gal}$$

$$Q = 2000 \text{ gpm} = \frac{2000}{7.48} = 267 \text{ ft}^3/\text{min}$$

$$d = 12 \text{ in} = 1 \text{ ft}$$

$$A = \frac{\pi d^2}{4} = \frac{\pi(1)^2}{4} = 0.785 \text{ ft}^2$$

$$V = \frac{Q}{A} = \frac{267 \text{ cfm}}{0.785 \text{ ft}^2} = 340 \text{ ft/min}$$

$$V = \frac{340}{60} = 5.7 \text{ ft/sec}$$

What is the velocity head for the flow given in Example 3–1?

Solution:

$$V = 5.7 \text{ ft/sec}$$

$$\text{velocity head} = \frac{V^2}{64.4} = \frac{(5.7)^2}{64.4} = 0.5 \text{ ft}$$

Example 3–2

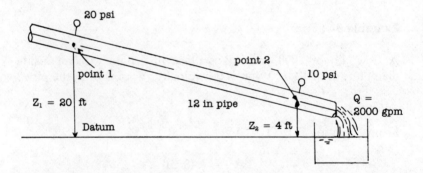

With conditions prevailing as indicated, what would be the total energy or head at point 1? (From Example 3–1, V = 5.7 ft/sec and $V^2/2g = 0.5$ ft/sec.)

$$H_1 = \left(\frac{p_1}{w}\right) + \frac{V_1^2}{2g} + Z_1$$

$$H_1 = \frac{20 \times 144}{62.4} + 0.5 + 20 = 66.65 \text{ ft}$$

What is the total energy at point 2, and how much energy is expended in moving 2000 gpm from point 1 to point 2?

$$H_2 = \frac{p_2}{w} + \frac{V_2^2}{2g} + Z_2$$

$$H_2 = \frac{10 \times 144}{62.4} + 0.5 + 4 = 27.58 \text{ ft}$$

$$(H_1 - H_2) = 66.65 - 27.58 = 30.07 \text{ ft}$$

Several basic principles are illustrated by the previous examples. First, the computation for a velocity head is usually complicated by the fact that quantity of flow may be in gallons. It must always be converted to cubic feet per second. The pipe area must always be in square feet. Velocity head then can be converted from $V^2/2g$ to $Q^2/A^2 2g$ using $Q/A = V$. Inserting the proper conversion units, 7.48 gal = 1 ft^2 and 60 sec = 1 min,

$$\frac{V^2}{2g} \text{ becomes } \frac{Q^2 4^2}{(60 \times 7.48)^2 \, \pi^2 d^4 2g}$$

$$\frac{V^2}{2g} = \frac{Q^2}{7.99 \times 10^6 d^4}$$

where Q = gpm
   d = diameter ft

If Q is expressed as discharge in thousands of gallons, then

$$\frac{V^2}{2g} = \frac{Q^2}{7.99 d^4}$$

Example 3–3

A pump is discharging 2000 gpm through a 12-in pipe. What is the velocity head developed?

$$\frac{V^2}{2g} = \frac{Q^2}{7.99d^4} = \frac{(2)^2}{7.99(1)^4} = 0.5 \text{ ft}$$

Example 3–4

If the flow is not as much as 1000 gpm, the expression works just as well. A pump discharges 200 gpm through a 4-in pipe. What is the velocity head developed?

$$\frac{V^2}{2g} = \frac{Q^2}{7.99d^4} = \frac{(0.2)^2}{7.99\left(\dfrac{4}{12}\right)^4}$$

$$\frac{V^2}{2g} = \frac{0.04}{(7.99)(0.0123)} = 0.407 \text{ ft}$$

The second fact in this general discussion of energy which should be noted is that the magnitude of the velocity head is relatively small in contrast to the other factors making up the total head.

Velocities in pipelines or sewers rarely exceed 10 fps, and most of the time velocities are 2–5 fps. This makes the velocity head a small part of the total energy in most cases.

$$V = 2 \text{ ft/sec} \quad \frac{V^2}{2g} = 0.062 \text{ ft}$$

$$V = 5 \text{ ft/sec} \quad \frac{V^2}{2g} = 0.388 \text{ ft}$$

$$V = 10 \text{ ft/sec} \quad \frac{V^2}{2g} = 1.55 \text{ ft}$$

On this basis, a velocity head is frequently neglected when it becomes a small part of the total head (say, one or two percent of the total).

A further point to note is that for a pipe which maintains the same size, since Q = AV, obviously the velocity is the same from point to point and thus the velocity head is the same. All this depends on the fact that no flow is taken out of the system between the points and that the flow is not varied.

Frequently, then, velocity heads can be neglected without any error, purely because they are the same from point to point and can be canceled out of any equation.

Finally, when thinking of a pipe obtaining water from any source, the usual situation would be to have the pipe fairly level (as in a water distribution system in fairly level ground). In this case, elevation differences become small also. In the last analysis, therefore, while the student always watches the three forms of head, the most important factors in pipe flow are generally pressure head, p/w_1, and head loss which will be discussed later.

All of these remarks apply strictly to pipe flow, of course. Flow in open channels is a special case where p/w forces are zero. This type of problem is taken up in Chapter 8.

PROBLEMS

3–1.   A 24-in diameter pipeline carries water at a velocity of 12 ft/sec. At a section of the same line, the diameter is 3 ft. What is the velocity at this cross section?

Answer: V = 5.3 fps

3–2.   The velocity of flow in a pipe with a diameter of 1 ft is 100 ft/min. The flow discharges through a nozzle, 3-in in diameter, attached to the end of the pipe. What is the velocity of flow at the tip of the 3-in nozzle in ft/sec?

Answer: 26.7 fps

3–3.   A flow of 1 mgd occurs in an 8-in pipeline. What is the flow velocity?

Answer: 4.44 fps

3-4. For a flow of 70.7 cfs and a velocity of 10 fps in a pipe, compute the inside diameter.

Answer: 3 feet

3-5. A flow of 1000 gpm passes through a 12-in diameter pipe which later reduces to 6 in in diameter. Calculate the velocities in the two sections of pipe.

Answer: $V_{12} = 2.8$ fps
$V_6 = 11.2$ fps

3-6. Water flows under pressure in a 4-in diameter pipe at a velocity of 10 ft/sec. If the pressure in the pipe at a point 5 feet above a reference elevation or datum is 500 lb/ft^2, what is the total head at that point?

Answer: 14.6 ft

3-7. The total head in a 6-in diameter pipe at a point 13.5 ft above a datum is 19.3 ft. If the pressure at that point is 312 lb/ft^2, what is the velocity of flow? What is the flowrate?

Answer: $V = 7.18$ fps
$Q = 1.41$ cfs

3-8. What is the velocity in a 6-in diameter pipe connected to a 1-ft diameter pipe if the flow in the larger of the two pipes is 1.42 cfs?

Answer: 7.24 ft/sec

4

Fundamentals of
Flow II

THE BERNOULLI EQUATION

Various hydraulic problems can be solved by comparing total energy or head between different points in a system. As noted in the last chapter, energy is required to move water from one point to another. An equality can be arranged which results in one of the most vital equations in hydraulics (Equation 4.1). Together with the equation of continuity, $Q = AV$, many hydraulic problems can be solved. The equality referred to is Bernoulli's equation and it consists entirely of a total energy (or head) relationship between successive points in a continuous steady state system. The term *steady state* means that flow conditions are constant.

$$(\text{Energy})_1 = (\text{Energy})_2 + (\text{Energy loss})_{(1 \text{ to } 2)} \qquad (4.1)$$

This can be written in other ways also. For example, it might be stated:

$$(\text{Total head})_1 = (\text{Total head})_2 + (\text{Head loss})_{(1 \text{ to } 2)}$$

It is most frequently stated by using the nomenclature developed in previous chapters:

$$\frac{p_1}{w} + \frac{V_1^2}{2g} + Z_1 = \frac{p_2}{w} + \frac{V_2^2}{2g} + Z_2 + (H_L)_{(1-2)} \qquad (4.2)$$

where Z_1 = elevation above datum at 1
   Z_2 = elevation above datum at 2
   H_L = head loss from 1 to 2

All other values have been defined in previous chapters.

It must be reemphasized that all the above terms must be in feet of head. They can then be added or subtracted as necessary to solve for the term in the equation which is of interest to the student.

The term (H_L), called *head loss,* is also, of course, measured in feet when Bernoulli's equation is used. Many items contribute to the head loss. In this lesson, all such items will be considered as being lumped together forming one figure of loss. Subsequent chapters will break the head loss into its component parts.

Two areas of head loss must be considered. The two areas are (1) losses due to the flow through a pipe or channel and (2) losses due to all other causes, such as bends, enlargements, obstructions, etc. These will be developed in later chapters.

As can be noted, the head loss is really energy loss due to friction. The first category above is normal friction induced by the fluid rubbing against the conduit. The second category involves loss because internal friction is increased between the water particles themselves. The bend, enlargement, or obstruction, etc., causes additional turbulence to that normally existing, and this results in added internal friction.

USE OF BERNOULLI'S EQUATION

In application, Bernoulli's equation is essentially simple. The student is cautioned to write each term whether or not it is zero since mistakes are made because terms are omitted. Then each term should be inserted into the equation either as a known quantity or as zero. Solution of the unknown factors is usually made in combination with the equation of continuity, $Q = AV$.

There are two factors which require attention. The choice of points 1 and 2, and the choice of a location for the datum plane.

Proper selection of these two items greatly assists in solving complex problems.

A further point must be emphasized. With one equation a student can solve for one unknown factor. Even with Q = AV and Bernoulli's equation only two unknown factors can be solved for. This should not stop one from trying to solve the problem when there are several unknown values. Frequently, choices of points or datum planes can cause factors to cancel out of the equation. A velocity head may be assumed small and neglected, if it would permit an approximate solution for an answer.

Example 4–1

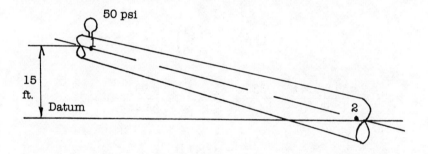

A pipeline gradually enlarges from 24 in in diameter at point 1 to 36 in in diameter at point 2. The velocity at 1 is 5 ft/sec and the average pressure at that same point is 50 psi. If all losses are lumped together and found to be 2 ft, what will the pressure be at point 2 if the pipe slopes 15 ft downward from 1 to 2?

First, assign points and place datum at point 2. Then write Bernoulli's equation from 1 to 2.

$$\frac{p_1}{w} + \frac{V_1^2}{2g} + Z_1 = \frac{p_2}{w} + \frac{V_2^2}{2g} + Z_2 + (H_L)_{(1-2)}$$

Examine each item in the equation:

$$p_1 = 50 \text{ psi}$$

$$V_1 = 5 \text{ ft/sec}$$

$$Z_1 = 15 \text{ ft}$$

$$Z_2 = 0$$

$$(H_L)_{(1-2)} = 2 \text{ ft}$$

There are still two unknown factors, namely p_2/w and $V_2^2/2g$. However, $A_1V_1 = A_2V_2$ will give the value for V_2 and the solution for p_2/w would be straightforward from that point on. Substitution into the continuity and Bernoulli equations gives:

$$A_1V_1 = A_2V_2 \text{ and } V_2 = \left(\frac{24}{36}\right)^2 \times 5 = 2.22 \text{ fps}$$

$$\frac{50 \times 144}{62.4} + \frac{25}{64.4} + 15 = \frac{p_2}{w} + \frac{4.93}{64.4} + 0 + 2$$

$$\frac{p_2}{w} = 128.4 \text{ ft}$$

$$p_2 = 128.4 \text{ ft} \times 62.4 \text{ lb/ft}^3$$

$$p_2 = 8025 \text{ lb/ft}^2$$

$$p_2 = 8025 \text{ lb/ft}^2 \times \frac{1 \text{ ft}^2}{144 \text{ in}^2}$$

$$p_2 = 55.7 \text{ psi}$$

It must be noted by the student that a methodical approach to the solution of the problem usually provides the framework for a logical answer. The following problem illustrates a different unknown.

Example 4-2

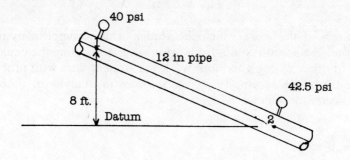

A 12-in pipe discharges water at the rate of 5.5 cfs. At point 1 on the pipe, the pressure is 40 psi, while at point 2 where the pipe is 8 ft lower the pressure is 42.5 psi. Compute the head loss from 1 to 2.

In the same fashion as Example 4-1, the facts are arranged.

$$\frac{p_1}{w} + \frac{V_1^2}{2g} + Z_1 = \frac{p_2}{w} + \frac{V_2^2}{2g} + Z_2 + (H_L)_{(1-2)}$$

$$p_1 = 40 \text{ psi}$$

$$Z_1 = 8 \text{ ft}$$

$$p_2 = 42.5 \text{ psi}$$

$$Z_2 = 0$$

Consideration of the known factors shows that 3 are unknown, but that since the pipe diameter, d, is known and Q is given as 5.5 ft³/sec, V_1 and V_2 could be found from Q = AV. There is no need to do this, however, since the pipe size is constant, thus $V_1^2/2g$ is equal to $V_2^2/2g$ and they cancel out of the initial equation. Therefore:

$$\frac{40 \times 144}{62.4} + 8 = \frac{42.5 \times 144}{62.4} + 0 + H_L$$

$$H_L = 92.3 + 8 - 98.1$$

$$H_L = 2.2 \text{ ft}$$

Much of the work which the student can accomplish using Bernoulli's equation comes from prior experience using the equation. In other words, the more work the student does with problems, the more accomplished he becomes in analysis of more complicated problems.

Example 4–3

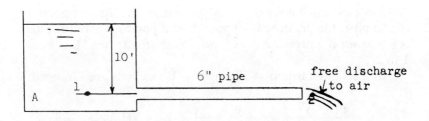

Assuming reservoir A to be so large that flow at point 2 is steady, the velocity at point 1 is zero, and the $(H_L)_{(1-2)}$ is 6 ft, what would the discharge at 2 be?

$$\frac{p_1}{w} + \frac{V_1^2}{2g} + Z_1 = \frac{p_2}{w} + \frac{V_2^2}{2g} + Z_2 + (H_L)_{(1-2)}$$

If a datum plane can be passed through points 1 and 2, then the following is true:

$$\frac{p_1}{w} = 10 \text{ ft}$$

$$\frac{p_2}{w} = 0, \text{ since free discharge}$$

$$10 + 0 + 0 = 0 + \frac{V_2^2}{2g} + 0 + 6$$

$$\frac{V_2^2}{2g} = 10 - 6 = 4$$

$$V_2 = \sqrt{4 \times 64.4} = 16.06 \text{ fps}$$

$$Q_2 = 0.196 \times 16.06 = 3.14 \text{ cfs}$$

Finally, the student must gain some ability to recognize false answers when they are produced.

Example 4–4

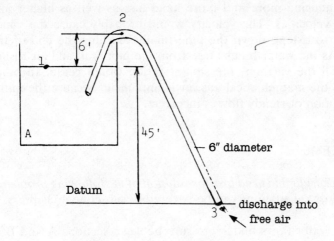

If steady-state conditions out of the reservoir prevail, what pressure exists at the top of the siphon shown when $(H_L)_{(1-2)}$ is 4 ft and $(H_L)_{(2-3)}$ is 8 ft?

$$\frac{p_2}{w} + \frac{V_2^2}{2g} + Z_2 = \frac{p_3}{w} + \frac{V_3^2}{2g} + Z_3 + (H_L)_{(2-3)}$$

$$\frac{p_2}{w} + \frac{V_2^2}{2g} + 51 = 0 + \frac{V_3^2}{2g} + 0 + 8$$

since $V_2^2/2g = V_3^2/2g$, they cancel.

$$\frac{p_2}{w} = 8 - 51$$

$$\frac{p_2}{w} = -43 \text{ ft of water}$$

In this case, the problem is set up correctly, but the answer is impossible since the maximum negative theoretical pressure is atmospheric pressure, which is 34 ft of water. Therefore, the answer is unsatisfactory. The answer is that as the negative pressure or vacuum builds up at the summit, the liquid will pull away from the pipe wall and the pipe will flow less than full. A pocket of water vapor will fill the pipe, restricting the cross section of the pipe, inducing more and more head loss as well as higher and higher velocities. The velocity would probably cause the vapor pocket to extend down the pipe finally reaching the end of the pipe. As the water breaks free from the pipe wall and air rushes in to fill the vacuum, the siphon action would cease. In other words, the mathematical answer is unrealistic because the initial assumption of steady flow is incorrect.

PROBLEMS

Assume negligible head loss for solution of the following problems. Head loss calculations will be covered in subsequent chapters.

4-1. Water flows under pressure between sections A and B in a pipeline. The diameter, d_a, at section A is 1 ft, and the diameter, d_b, at section B is 2 ft. Section A is 10 ft above the datum and section B is 25 ft above the datum. The velocity, V_a, at section A is 16.8 fps and at B, V_b, is 4.2 fps. The pressure p_a, at section A is 1379 lb/ft². (a) Compute the pressure at section B in pounds per square foot. (b) What is the pressure head at section B? (c) What is the total head at section A? At B?

Answer: a. p_b = 700 psf
   b. $\dfrac{p_b}{62.4}$ = 11.2 ft
   c. 36.5 ft at both sections.

4–2.   A horizontal 6-in diameter pipe carries water at a pressure of 60 psi. The pipe reduces to a horizontal 3-in diameter section in which the pressure is 20 psi. What is the flow through the two pipes?

Answer: 3.9 cfs

4–3.   A 12-in pipe contains a short section where the diameter is reduced to 6 in, and then enlarged again to 12 in. A point in the 6-in section is 2 ft below a point in the 12-in line at which the pressure is 10,800 psf. If a flow of water through the line is 4.25 cfs, what is the pressure head at the point in the 6-in line? What is the pressure at that point in lb/ft²?

Answer: Pressure head = 168.3 ft
            p_6 = 10,498 psf

4–4.   A 20-ft diameter horizontal pipe has a reducing section containing a 16-ft diameter pipe. The pressure in the 20-ft diameter pipe is 120 psi, and the velocity there is 50 fps. What are the values for the pressure and velocity in the 16-ft diameter pipe?

Answer: V_{16} = 78.1 fps
            p_{16} = 95.7 psi

4–5.   Calculate the velocity, V_1, in a 3-in diameter fire hose discharging 0.70 cfs. If the flow discharges through a 1.25-in diameter nozzle, calculate the pressure intensity, p_1, in the 3-in hose just upstream from the nozzle, and calculate the velocity, V_2, of discharge from the nozzle. (Note: p_2 = 0, since free discharge.)

Answer: V_1 = 14.3 fps
            p_1 = 44.2 psi
            V_2 = 82.3 fps

5

Pipe Flow with Friction Loss

As noted in Chapter 3, pipe flow occurs whenever a moving fluid completely fills a closed conduit of any shape. The principles of pipe flow are applicable generally to the hydraulics of such structures as pressurized pipe systems, submerged inlets to basins, siphons, and various other types of pipelines.

Because pipe flow occurs in a closed conduit, the cross sectional area of the flow is fixed by the cross section of the conduit and, of course, the water surface is not open to the atmosphere. Thus, for this flow condition, the pressure in the conduit may be equal to, greater than, or less than atmospheric pressure. For pipe flow, the cross sectional area of the flow as well as the hydraulic radius remains constant for any particular cross section of the conduit. For the common case in which the total discharge or flow of water is constant, the velocity by any cross section is likewise constant. If the flow should vary, the velocity would be directly proportional to the discharge or flowrate.

In the great majority of cases encountered in the water and wastewater field, pipe flow occurs at large enough velocities to make the flow turbulent.

FRICTION LOSS

For the ideal case, water flowing in a pipe would encounter no resistance other than that due to pressure differences. In practical application, however, turbulence exists at the boundary between the sides of the conduit and the flowing water (see Figure 3–2), resulting in a certain loss of energy or head. This loss of head is due to the friction created between the fluid and the wall of the conduit, and is appropriately termed the *friction loss*. As noted in Chapter 4, the friction loss, H_L, must be added to the Bernoulli equation, giving the form shown in Equation 4.2, which is repeated here:

$$\frac{p_1}{w} + \frac{V_1^2}{2g} + Z_1 = \frac{p_2}{w} + \frac{V_2^2}{2g} + Z_2 + (H_L)_{(1-2)}$$

HYDRAULIC GRADE LINE

If a vertical glass tubing were inserted into the side of a pipe, the water would rise in the tube due to the water pressure in the pipe. The water would rise to an elevation equivalent to the pressure, in feet. This value is represented in the Bernoulli equation by p/w. If several glass tubes were inserted in the pipe, and a line were drawn connecting the points to which the water rises, this line would represent the hydraulic grade line. For pipe flow in a level straight conduit of constant cross section and uniform wall roughness, the rate of loss of head due to friction is constant, and the hydraulic grade line in the direction of flow along the conduit has a slope equal to the friction head loss per foot of conduit. Thus, the hydraulic grade line is actually a plot of the term p/w above the centerline of the pipe or, more commonly, is a plot of $(p/w + Z)$ above the datum along the pipe.

The energy grade line, on the other hand, is a line joining a series of points representing the *total* energy along the length of the pipe. Thus, the energy grade line consists of a plot of $p/w + Z + V^2/2g$ above the datum along the pipe. By definition, then, the energy grade line is always higher than the hydraulic grade line by a vertical distance of $V^2/2g$ at each point.

An example of hydraulic and energy grade lines is given in Figure 5–1 for a pipeline flowing under pressure.

It may be seen in Figure 5–1 that the hydraulic and energy grade lines at point 2 on the pipe are lower than the corresponding grade lines at point 1. The difference in the grade line at points

1 and 2 represents the losses which occur between these two points. As has been noted, the friction developed against the walls of the conduit and within the water itself results in losses of head which are related to the length of the pipe, its size, its condition (smoothness or roughness), and the velocity of flow. In Figure 5–1, the head loss between points 1 and 2—difference in the heights of the grade lines—has been consumed in producing the flow. In other words, pressure is reduced in order to cause flow.

As mentioned in Chapter 4, in most problems in fluid flow, friction losses may be divided into two categories (1) friction losses in lengths of pipe and (2) friction losses in bends, valves, obstructions, changes in section, etc. It is common practice to refer to the friction losses in pipes as *major losses,* and those in bends, valves, etc., as *minor losses.* In this chapter, pipe losses, or major losses, will be considered, while minor losses will be discussed in Chapter 7. With all of these terms considered, the Bernoulli equation can be written:

$$\frac{p_1}{w} + \frac{V_1^2}{2g} + Z_1 = \frac{p_2}{w} + \frac{V_2^2}{2g} + Z_2 + H_{L(Pipe)} + H_{L(Minor)} \quad (5.1)$$

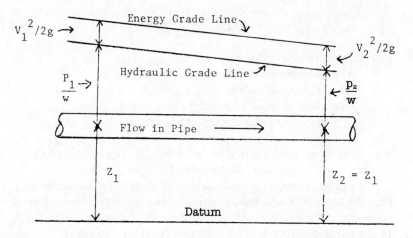

Figure 5–1. Hydraulic and energy grade lines.

CALCULATION OF HEAD LOSS DUE TO FRICTION

Head loss in a pipe can most accurately be calculated using the pipe-friction equation (sometimes called the Darcey-Weisbach equation).

$$H_L = f \frac{L}{d} \frac{V^2}{2g} \qquad (5.2)$$

where H_L = total head loss along pipe of length L, ft
   f = friction factor; relates to the degree of roughness of the pipe (obtained using a Moody diagram)
   L = length of pipe, ft
   d = diameter, ft
   V = velocity, ft/sec
   g = gravitational constant, ft/sec^2

This equation requires the use of a Moody diagram (which can be found in most hydraulic handbooks) for determination of the friction factor (f). A more commonly used equation is the Hazen-Williams equation:

$$V = 1.318 \, CR^{0.63}S^{0.54} \qquad (5.3)$$

where V = velocity, ft/sec
   C = roughness coefficient (C factor)
   R = hydraulic radius, ft
   S = feet of head loss per foot of length (slope of the energy grade line)

The roughness coefficient C is commonly referred to as the C factor. Some common C factors are given in Table 5–1.

It will be recalled (from equation of continuity, Equation 3.1), that the total flow in a pipe is equal to the velocity times the cross sectional area, or Q = AV. Therefore, if Q is substituted for V in Equation 5.3, the Hazen-Williams formula becomes:

$$Q = 1.318 \, ACR^{0.63}S^{0.54} \qquad (5.4)$$

Table 5-1 Approximate Values for Hazen-Williams Roughness Coefficient for Various Types of Pipes

Pipe Description	C Factor
Asbestos cement	140
Brass or lead, new	140
Ductile iron, cement lined	140
Cast iron, uncoated	
10 years old	110
15 years old	100
20 years old	90
30 years old	80
Concrete	
Very smooth, excellent joints	140
Smooth, good joints	120
Rough	110
Corrugated	60
Steel, welded, new	130
Vitrified	110
Wrought iron or standard galvanized steel	
Diameter 12 and larger	110
Diameter 4 to 12	100
Diameter 4 and smaller	80

As we saw before, for round pipes flowing full the hydraulic radius (R) is equal to the inside diameter divided by four (R = d/4), and the cross sectional area of the pipe is equal to π times the square of the diameter divided by four (A = $\pi d^2/4$). Therefore, the inside diameter can be substituted for both the hydraulic radius and the area terms in Equation 5.4 to give:

$$Q = 0.432 \ Cd^{2.63}S^{0.54} \qquad (5.5)$$

An alignment chart for solution of the Hazen-Williams equation is given in Figure 5-2. This chart makes solution possible for head loss problems by the Hazen-Williams equation without cal-

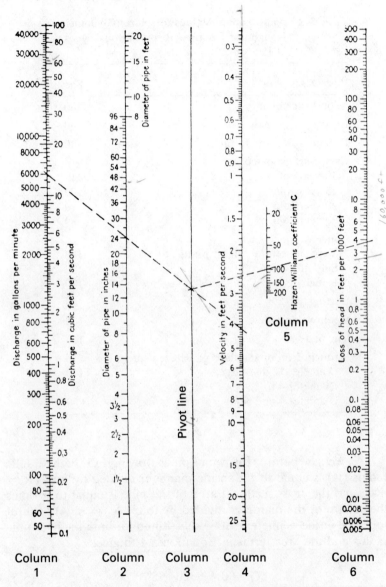

Figure 5–2. Hazen-Williams alignment chart.

culating the 2.63 and 0.54 powers of d and S, respectively. A direct solution is possible if a calculator with a power function is available.

DIRECTIONS FOR USING
THE HAZEN-WILLIAMS
ALIGNMENT CHART

For determining the head loss (Column 6), find the proper discharge in Column 1 and line this up with the diameter of the pipe in Column 2. Extend this line across the pivot line in Column 3, read the velocity in Column 4, and mark the point where the pivot line was crossed. Line up the point that was marked on the pivot line with the proper Hazen-Williams coefficient C, extend this line to Column 6, and read the head loss per 1000 feet. Columns 4, 5, and 6 are always lined up with the crossing point on the pivot line. The velocity (Column 4) is always determined from the line going through Columns 1 and 2. The following examples demonstrate the use of the chart.

Example 5–1

A 24-in, 15-year-old, uncoated cast iron pipe carries a total flow of 6000 gpm. (a) What is the velocity of flow in the pipe? (b) What is the head loss due to friction per 1000 ft of the pipe? (c) What is the value of S (head loss per foot)? (d) If the pipe is 10,000 ft long, what is the total head loss?

Solution:

From Table 15–1, a 15-year-old uncoated cast iron pipe has a Hazen-Williams coefficient of C = 100. Now, in Figure 5–2, draw a straight line connecting 6000 gpm on the discharge (Q) scale (Column 1) and 24 in on the diameter scale (Column 2).

Extend this line to the velocity scale and read off velocity. Mark the point where this line crosses the pivot line (Column 3). Now connect this point on the pivot line with the proper value on the C scale (Column 5) and extend across to the loss of head line (Column 6). Read from the head loss scale the value for head loss where the line crosses that scale.

a.   From Figure 5–2, V = 4.2 fps
b.   From Figure 5–2, head loss/1000 ft = 3.9 ft/1000 ft
c.   S = head loss/ft = 3.9/1000 = 0.0039 ft/ft
d.   If the pipe is 10,000 ft long, the total head loss is

$$(H_L) = \frac{0.0039 \text{ ft}}{\text{ft}} \times 10,000 = 39 \text{ ft}$$

Example 5–2

A new asbestos cement pipe must be selected to carry a flow of
5 cubic feet per second (cfs). (a) If the head loss in the pipe due
to friction is not to exceed 10 ft/1000 ft, what is the smallest
diameter pipe that can be used? (b) What will be the velocity of
flow in the pipe?

Solution:

From Table 5–1, a new asbestos cement pipe has a C value of
140. In Figure 5–2, draw a line connecting C = 140 (Column 5)
and the head loss value of 10 ft/1000 ft (Column 6), and extend
the line to cross the pivot line (Column 3). Connect that point on
the pivot line to a value of 5 cfs on the Q scale (Column 1).

a.   From Figure 5–2, d = 12 in
b.   From Figure 5–2, V = 6.5 fps

Example 5–3

What is the value of C for an old 10-in pipe which is found to
carry a flow of 2000 gpm with a head loss of 110 ft/1000 ft?

Solution:

Connect 2000 on the Q scale (Column 1) in Figure 5–2 with 10
in on the d scale (Column 2). Extend to cross pivot line (Column
3). Connect point on pivot line with 110 on head loss scale (Column 6). Read a value of 55 from the point where the line crosses
the C scale (Column 5). From the velocity scale, we can also
determine that V = 8.2 fps.

CLASSIFICATION OF PROBLEMS

In any practical problem in hydraulics, one or more of the important parameters previously discussed will be known and others will have to be determined. In this connection, a classification of types of problems appears appropriate at this time.

Whenever any two of the three related parameters, A, d, or V are known, and assuming a value for C, the third can readily be obtained either by direct calculation or graphically from a chart. In such a case, the flow regime is rigorously described, dictating a definite hydraulic grade line or slope to keep the water flowing. This slope and, consequently, the head loss, may thus be obtained directly from Figure 5–2.

If the pipe diameter, length, available head, and C factor are known and the velocity and head loss are unknown, the problem is somewhat more complex. A Bernoulli equation made between two properly selected points in the system will show that the total head loss is equal to the drop in pressure plus the elevation difference if the velocity head $V^2/2g$ can be neglected. (See Example 3–4.) By this method, flow conditions can be approximated.

To clarify these points, sample problems with solutions will be shown.

Example 5–4

Flow from tank to tank:

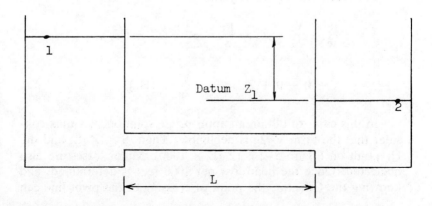

Bernoulli's equation in this case shows that the total head loss between points 1 and 2 is exactly equal to the value of Z_1, the elevation difference, since pressures at both ends are zero. Velocities at both ends are also zero.

$$\frac{p_1}{w} + \frac{V_1^2}{2g} + Z_1 = \frac{p_2}{w} + \frac{V_2^2}{2g} + Z_2 + (H_L)_{(1-2)}$$

$$0 + 0 + Z_1 = 0 + 0 + 0 + (H_L)_{(1-2)}$$

Hence, the value of S in the Hazen-Williams equation (neglecting minor losses in bends, entrances, etc.) is equal to Z_1/L, and the H_L point on Figure 5–2 would be $(Z_1/L) \times 1000$.

Example 5–5

Flow from tank with open-end discharge:

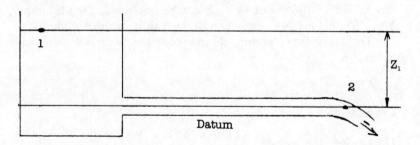

The Bernoulli equation yields:

$$\frac{p_1}{w} + \frac{V_1^2}{2g} + Z_1 = \frac{p_2}{w} + \frac{V_2^2}{2g} + Z_2 + (H_L)_{(1-2)}$$

$$0 + 0 + Z_1 = 0 + \frac{V_2^2}{2g} + (H_L)_{(1-2)}$$

In this case, to obtain an appropriate solution, we must consider that the term $V_2^2/2g$ is negligible. Then, $S = Z_1/L$, and the H_L point on Figure 5–2 is $(Z_1/L \times 1000)$. Minor losses are also neglected. Once the head loss per 1000 feet is determined, and knowing the C factor, the point of crossing of the pivot line can be determined on Figure 5–2. Lining this point up with the pipe diameter, the velocity can then be determined. Once the velocity is determined, the term $V_2^2/2g$ should be calculated. If it is much less than $(H_L)_{(1-2)}$, the assumption that it was negligible was correct. If not, recalculate using the determined value of V_2 and solve for a new value of V_2. When the old and new values are close, the solution is complete.

Example 5–6

Straight pipeline flow with pressure gauges at points 1 and 2:

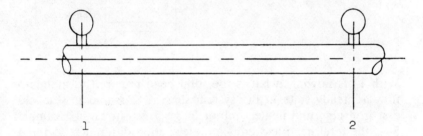

Write Bernoulli's equation between 1 and 2.

$$\frac{p_1}{w} + \frac{V_1^2}{2g} + Z_1 = \frac{p_2}{w} + \frac{V_2^2}{2g} + Z_2 + (H_L)_{(1-2)}$$

Z_1 and Z_2 are both equal to zero in this case and, if the pipe size remains constant from one point to the other, the terms $V_1^2/2g$ and $V_2^2/2g$ cancel, yielding:

$$(H_L)_{(1-2)} = \frac{p_1}{w} - \frac{p_2}{w}$$

which is the drop in pressure head.

If the pipe size changes, we can still neglect the velocity terms because their difference is likely to be small.

One further type of pipe-to-pipe situation must be considered; that in which a difference in pipe elevation exists. If the elevation difference is significant, the expression for head loss will have the form:

$$(H_L)_{(1-2)} = \frac{p_1}{w} - \frac{p_2}{w} + (Z_1 - Z_2)$$

which is the change in pressure plus the elevation difference.

Example 5–7

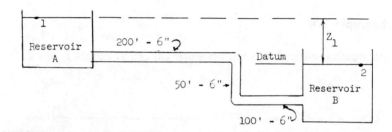

With 1 cfs flowing, what is the total head required to maintain flow at steady state in the system shown? (Assume 15-year-old cast iron pipe and neglect minor losses.) As shown in Example 5–4, the total head loss equals the elevation difference (Z_1) and is the head required to maintain the desired flow.

$$Q = 1 \text{ cfs}$$

$$d = 6 \text{ in}$$

Now, from Figure 5–2:

1. knowing Q and d, read V = 5 ft/sec;
2. establish turning point on chart;
3. using C = 100 from Table 5–1, read H_L = 24 ft/1000 ft;
4. since the system only contains 350 ft of 6-in pipe, the head loss from point 1 to point 2 is:

$$(H_L)_{(1-2)} = 24 \times 350/1000 = 8.4 \text{ ft}$$

The problem is not finished until it is certain that no items of significance have been neglected. Minor losses have been neglected in this problem, and if these are important they further complicate the problem. (This is discussed in detail in Chapter 7.)

Example 5–8

Referring to the sketch used in Example 5–7, assume the difference in elevation between reservoirs A and B is 13.4 ft; what is the flow that can be maintained?

The answer can be obtained readily. As in Example 5–4, Bernoulli's equation can be written between 1 and 2, showing that $(H_L)_{(1-2)} = Z_1$.

$$H_L = 13.4 \times \frac{1000}{350} = 38.3 \text{ ft per 1000 ft}$$

With $H_L = 38.3$ ft per 1000, and C = 100, find turning point in Figure 5–2; using turning point and 6-in diameter, read Q = 1.25 cfs.

Now if reservoir B is eliminated from the piping system, how should one approach the solution? The total elevation difference used, water surface to water surface, will still be 13.4 ft, as before. The following example shows the solution.

Example 5–9

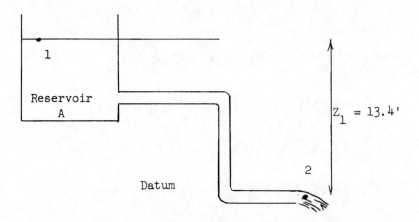

After drawing a good sketch, the first thing that strikes the eye is that the (Z_1) quantity can be much larger than the value of 13.4 ft in the former example, depending on the depth of connection to reservoir B in that example. This added head will tend to increase flow because the system does not have to work against the back pressure of some depth of water over the exit end of the pipe.

In this case, the assumption will be made that the pipe in the previous example discharged into reservoir B virtually at its surface so that little difference exists between the former Z_1 and the present Z_1.

Set up Bernoulli's equation and neglect $V_2^2/2g$ (as in Example 5–5). Also minor losses are neglected.

$$(H_L)_{(1-2)} = 13.4 \text{ ft}$$

$$H_L = 13.4 \times \frac{1000}{350} = 38.3 \text{ ft per 1000 ft}$$

As before, Q = 1.25 cfs.

The difference between this solution and the former (Example 5–8) is that velocity head has been neglected. It is apparent now that the significance of doing this can be determined. With the approximate value of Q being 1.25 cfs and diameter of 6 in, V = 6.4 fps, and going back to Bernoulli's equation:

$$0 + 0 + Z_1 = 0 + \frac{V_2^2}{2g} + 0 + (H_L)_{(1-2)}$$

$$(H_L)_{(1-2)} = 13.4 - \frac{V_2^2}{2g}$$

$$(H_L)_{(1-2)} = 13.4 - \frac{(6.4)^2}{64.4} = 13.4 - 0.61$$

$$(H_L)_{(1-2)} = 12.8 \text{ ft}$$

$$H_L = 12.8 \times \frac{1000}{350} = 36.6 \text{ ft per 1000 ft}$$

New Q = 1.20 cfs

This is not too different from 1.25 cfs, and shows that $V_2^2/2g$ was indeed negligible.

(Note: If the values for Q were calculated using the Hazen-Williams equation instead of the chart, the first value for Q would have been 1.20 cfs instead of 1.25 cfs and the new Q would have been 1.17 cfs instead of 1.20 cfs.)

PROBLEMS

5–1.   Water flows upward at a rate of 7.85 cfs in a vertical 12-in diameter pipe. At point a in the pipe the pressure, P_a, is 4392 psf. At point b, 15 ft above point a, the pipe diameter is 24 in, and pressure, P_b, is 3082 psf. What is the friction loss in head between points a and b?

   Answer: $(H_L)_{(a-b)} = 7.5$ ft

5–2.   A flow of 10 cfs occurs in a rough concrete pipe of 30-in diameter. What is the velocity of flow? Calculate the total head loss due to friction in a 2000-ft length of the pipe.

   Answer: $V = 2.04$ fps
   $(H_L) = 1.3$ ft

5–3.   Determine the value of the Hazen-Williams coefficient, C, for a 2-ft diameter pipe carrying 20 cfs with a total head loss of 50 ft over a 5000-ft length.

   Answer: $C = 90$

5–4.   An upward inclined 6-in pipe discharges water into the atmosphere. The flow rate is 1.963 cfs and the pressure at a point in the pipe 6 ft below the discharge end is 33.5 psi. Calculate the total head loss between that point and the discharge end.

   Answer: $(H_L) = 71.4$ ft

5–5.   What minimum size vitrified pipe should be used to carry a flow of 17 cfs if the total friction loss is to be 25 ft over a pipe length of 5000 ft? What is the velocity of the flow?

   Answer: $d = 24$ in
   $V = 5.4$ fps

6

Compound Pipes

In many problems encountered in the analysis of hydraulic systems, one is concerned with pipes of different sizes connected end to end, or with two or more pipes of the same size or of different sizes connected in parallel. The existence of such systems necessitates some discussion of *pipes in series* (pipes of different sizes connected end to end), *pipes in parallel* (two or more pipes operating side by side, thus splitting the total flow between them), and equivalent pipes. When pipes are connected in series or parallel, they are said to comprise a *compound pipe* system. Examples of these situations are shown in Figure 6–1.

For solution of hydraulic problems related to compound pipe systems, a useful concept is that of *equivalent pipes*. Two pipes (or two pipe systems) are said to be equivalent if the losses of head in each are identical for the same rates of flow. The equiv-

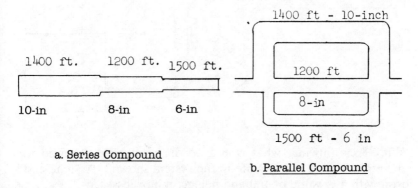

Figure 6–1. Compound pipe systems.

65

alent pipe concept consists of replacing several connected pipes (in series or parallel) with one pipe of constant diameter which creates the same total head loss for a given flow. The equivalent pipe length is determined by selecting a pipe length and diameter that will give the same head loss as the total head loss for the original multipipe system at a given or assumed flow.

PIPES IN SERIES

A compound system involving pipes in series may be approached in essentially the same fashion as a simpler system made up of sections of pipe of the same diameter. That is, the head losses for each section may be added to give the total head loss for the entire length of pipe. Example 6-1 demonstrates the proper approach to a problem of this type.

Example 6-1

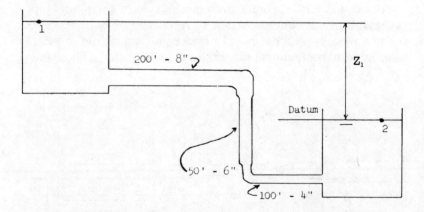

With 1 cfs flowing, what would be the total head required to maintain flow at steady state in the system shown? (Assume cast iron with a C value of 100 and neglect minor losses.)

From Figure 5–2, the following slopes are obtained:

8-in pipe at 1 cfs = 7 ft/1000 ft or 0.007 ft/ft
(head loss per foot of pipe)
6-in pipe at 1 cfs = 24 ft/1000 or 0.024 ft/ft
4-in pipe at 1 cfs = 200 ft/1000 or 0.200 ft/ft

Now, multiplying the slopes by the appropriate lengths of the pipes:

200 ft of 8-in at 0.007 ft/ft = 1.40 ft
50 ft of 6-in at 0.024 ft/ft = 1.24 ft
100 ft of 4-in at 0.200 ft/ft = 20.00 ft

$$(H_L)_{(1-2)} = \text{Total} = 22.64 \text{ ft}$$

The complete solution of the above problem again involves checking Bernoulli's equation to be sure no factors have been overlooked. It should be recognized that the four inch pipe in the previous problem is contributing a major portion of the total head loss of the system, out of proportion to that contributed by the other sections of pipe. A check of the velocity shows that for Q = 1.0 cfs and d = 4 in, V = 11.5 fps.

It has been noted previously that velocities of 2 to 5 fps are normal. Velocities greater than 10 fps are often obtained in waste water lines or fire hoses which are not in continuous use, but rarely can a system afford to expend the energy required to pump at these higher velocities on a continuous basis. In this case, replacing 100 ft of 4-in with 100 ft of 6-in would drop the total head loss from 22.64 ft to 5.0 ft, a major saving in energy or power. This should emphasize why recommendations for distribution systems often will not permit use of 4-in pipes.

More complex problems involving series compound pipes arise when no information is available regarding values for two of the terms Q, A, or V, but instead a pressure or head differential is revealed by the Bernoulli equation, suggesting the availability of a hydraulic slope. In such complex cases, solution is facilitated by the use of equivalent pipes.

EQUIVALENT PIPE DETERMINATIONS FOR PIPES IN SERIES

The use of the concept of equivalent pipes in solving series compound pipe problems involves the conversion of lengths of pipe
of different sizes into a hypothetical length of pipe of one size
which will give exactly the same total head loss for the same flow.
The diameter chosen for the hypothetical equivalent pipe need
not be the same as that of any of the pipes in the actual system.
Once chosen, however, the diameter must be used consistently
thereafter in application of the Bernoulli equation between any
two points in the system, just as though the hypothetical pipe
were actually present. By way of example, consider a flow of 1
cfs taking place in a 1000-ft length of 10-in pipe with a Hazen-
Williams C value of 100. What would be the equivalent length of
a 6-in pipe having the same C value? To answer this question one
must first compute the head loss which would occur in the 1000-
ft length of 10-in pipe. From Figure 5–2, it may be seen that a
flow of 1 cfs in the 10-in pipe produces a slope, s, of 2.2 ft per
1000 ft of length. From the same chart, a value of s = 24 ft per
1000 ft of length is obtained for a 1 cfs flow in a 6-in pipe. Thus,
for the 10-in pipe:

$$H_L = \frac{2.2 \text{ ft}}{1000 \text{ ft}} \times 1000 \text{ ft} = 2.2 \text{ ft}$$

For the 6-in pipe, S = 24 ft/1000 ft (0.024 ft/ft), and the equivalent
length of the 6-in pipe is

$$L_6 = \frac{H_L}{S} = \frac{2.2 \text{ ft}}{0.024 \text{ ft/ft}} = 92 \text{ ft}$$

Thus, in terms of total head loss for a flow of 1 cfs the 92-ft
length of 6-in pipe is *equivalent* to the 1000-ft length of 10-in pipe.
The two systems should thus be equivalent for any other flow,
and should produce the same head losses at these flows. To sub-
stantiate this, consider a different flow, say Q = 1.5 cfs. From
Figure 5–2, it may be seen that for this flow the slope is 52 ft/
1000 ft of length for the 6-in pipe. Hence,

$$(H_L)_6 = \frac{52 \text{ ft}}{1000 \text{ ft}} \times 92 \text{ ft} = 4.8 \text{ ft}$$

Again, from Figure 5–2, the value of S per 1000 ft for the 10-in pipe at a flow of 1.5 cfs is 4.7 ft/1000 ft and

$$(H_L)_{10} = \frac{4.7 \text{ ft}}{1000 \text{ ft}} \times 1000 \text{ ft} = 4.7 \text{ ft}$$

In the use of nomographs such as Figure 5–2 for obtaining values for the various hydraulic parameters, one may note that slight discrepancies often exist in checking problems such as that given above. Such discrepancies may result simply from distortion of the nomograph. By this time, however, the approximate nature of hydraulic computations has been emphasized sufficiently, so that a certain tolerance of error (up to perhaps 2–5%) should be expected and allowed for.

With knowledge of the manner in which equivalent pipes may be substituted for actual pipes in any system, it is now possible to consider more complex cases of compound series arrangements in which only one of the factors, Q, A, or V, is known along with the hydraulic slope. In setting forth an example for this case, pipe lengths will be kept short so that the same system can be employed later, in Chapter 7, to illustrate differences produced in head loss when minor losses are included in the considerations.

Example 6–2

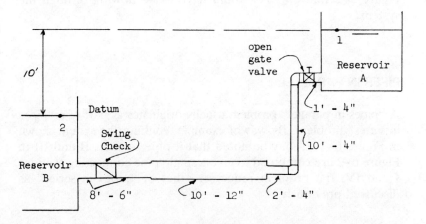

Assume that the system shown is flowing at steady state under the total head of 10 ft. What quantity of water would flow from reservoir A to reservoir B, assuming C = 100?

Neglecting minor losses, the Bernoulli equation is written:

$$\frac{p_1}{w} + \frac{V_1^2}{2g} + Z_1 = \frac{p_2}{w} + \frac{V_2^2}{2g} + Z_2 + (H_L)$$

$$0 + 0 + 10 = 0 + 0 + 0 + (H_L)$$

For equivalent pipe substitution, assume that 1 cfs is flowing through the system.

Pipe	H_L ft/1000	S (ft/ft)	Actual loss, ft
13 ft of 4-in	200.0	0.20	2.60
10 ft of 12-in	0.9	0.0009	0.0009
8 ft of 6-in	24.0	0.024	0.19
31 ft			2.79

Now, choosing 6-in pipe and knowing that S = 0.024 ft/ft for 1 cfs in such a pipe, then a pipe theoretically equivalent to the actual 31 ft of mixed pipe sizes would be 2.79/0.024 = 116 ft of 6-in pipe.

Merely insert 116 ft of 6-in pipe between the reservoirs, and the slope now is 10/116 = 0.086 ft/ft, or 86 ft/1000 ft. In order for such a loss of head to occur, one can now calculate from Figure 5–2 that 1.85 cfs would have to be flowing through the system.

PIPES IN PARALLEL

A "pipes in parallel" problem usually originates as part of a "pipes in series" problem. By way of example, consider the system shown in Figure 6–2. It may be noted that if pipe sections II and III in Figure 6–2 are combined to give a single pipe connecting sections I and IV, the problem reduces to that of "pipes in series" as discussed previously.

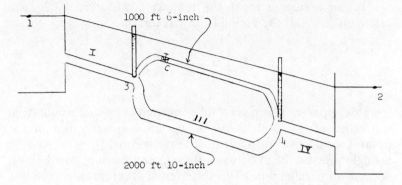

Figure 6–2. Parallel pipe system.

Continuing with the example, it is not difficult to realize that if the valve, C, in section II is closed, it becomes more difficult for water to flow from point 3 to point 4. In essence, one large pipe might also be used between points 3 and 4 which would be hydraulically equivalent to the parallel system for the case in which valve C is opened, but a smaller pipe would be required for hydraulic equivalency for the case in which valve C is closed. In other words, substitution of one equivalent pipe for the parallel system requires a matching of resistances to flow, or head losses.

In handling analyses of parallel pipe systems, it is invariably necessary to substitute one hydraulically equivalent pipe for the two or more pipes in parallel. By definition, this hypothetical pipe must pass the same flow with exactly the same head loss. It may have any diameter, but once d is chosen, the corresponding length to give hydraulic equivalency is fixed. Similarly, the hypothetical pipe could be chosen with a certain length, in which case its diameter would be uniquely fixed in order that it give the required head loss for the specific quantity of flow.

Referring once more to Figure 6–2, if the Bernoulli equation is written between points 3 and 4, and minor losses are neglected for the time being, one obtains

$$(H_L) = \left(\frac{V_3^2}{2g} - \frac{V_4^2}{2g}\right) + \left(\frac{p_3}{w} - \frac{p_4}{w}\right) + (Z_3 - Z_4)$$

Taking a case in which the velocity head terms, $V^2/2g$, are relatively low and the piping is on flat terrain,

$$\mathfrak{y}(H_L) = \left(\frac{p_3}{w} - \frac{p_4}{w}\right)$$

It is apparent that a glass tube inserted at point 3 would have water rising to (p_3/w) ft above the datum, and water in a tube at point 4 would rise to (p_4/w) ft. The differential $(p_3/w - p_4/w)$ is a single pressure head between these points causing flow in both sections of parallel pipe. This differential also represents the loss in head between points 3 and 4. Thus, both pipes have the same head loss and it becomes clear that one equivalent pipe can be substituted between points 3 and 4 yielding a head loss equal to $(p_3/w - p_4/w)$.

EQUIVALENT PIPE DETERMINATIONS FOR PIPES IN PARALLEL

In determining an appropriate equivalent pipe to connect sections 3 and 4 in Figure 6–2, one must first assume a value for the differential pressure, $(H_L)p$. Since this differential acts to force water through both sections, II and III, the next step is to determine what flow would be characteristic for each section at this head loss or a differential pressure. The solution of this problem is given in Example 6–3.

Example 6–3

Assume $(H_L) = 3$ ft, $C = 100$. Under this head, the discharge calculated from Figure 5–2 is:

$$\begin{aligned} \text{Pipe II,} \quad & Q = 0.3 \\ \text{Pipe III,} \quad & Q = \underline{0.8} \end{aligned}$$

$$\text{Total } Q = 1.1 \text{ cfs}$$

What single pipe would give the same total Q for (H_L) = 3 ft? (1) Choose diameter = 12 in. (2) Refer to Figure 5-2 and select turning point (for C = 100) for the given values and for Q and d. Then S = 1.1 ft/1000 ft = 0.0011 ft/ft.

Now (H_L) = S × L; dividing both sides of the equation by S gives:

$$L = \frac{(H_L)}{S}$$

$$L = \frac{3 \text{ ft}}{0.0011 \text{ ft/ft}} = 2730 \text{ ft}$$

Thus, a pipe of diameter d = 12 in and length L = 2730 ft is equivalent to the parallel sections II and III in Figure 6-2.

For further computation, the problem can now be considered as a "pipes in series" problem and treated in the same manner as discussed earlier in this chapter.

MORE COMPLEX SYSTEMS

Regardless of the complexity of a pipe system, the principles discussed previously may be applied for obtaining a correct hydraulic solution. Series systems of the highest degree of complexity may be reduced to a simple equivalent pipe of specific diameter and length, as long as the quantity of flow is considered to be at steady state.

For non-steady state flow, that is when water is drawn from the system at various points along the pipe, the approach one must take is to imagine that the piping is cut into sections at these points. Each section is then considered to be a separate problem with a different flow, and with the appropriate flow fixing the respective velocity for each section. The velocity in turn fixes the turning point on the Hazen-Williams chart and, if a roughness, C, is assumed then the corresponding hydraulic slope, S, can be read for each section of pipe.

As far as complex parallel systems are concerned, any number of parallel pipes connecting two points may be converted to one equivalent pipe between those points. When all parallel pipes have been substituted for by single equivalent pipes, the problem is handled as a "pipes in series" problem as above.

PROBLEMS

6–1. Convert the following pipes to equivalent lengths of 8-in pipe with C = 100. (a) 2000 ft of 12-in with C = 120; (b) 500 ft of 6-in with C = 120; (c) 200 ft of 6-in with C = 120.

Answer: Equivalent lengths of 8-in with C = 100 are: (a) 200 ft; (b) 1500 ft; (c) 600 ft.

6–2. Using the equivalent pipe method, determine the flow in the series compound pipe shown below if pressure at point 1 is 100 psi and that at point 2 is 50 psi. Assume C = 120 for all pipes.

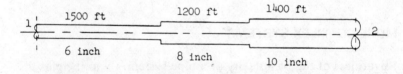

Answer: Q = 1.85 cfs

6–3. The flow from point 1 to point 2 in the parallel pipe system shown below is 4.45 cfs. What is the head loss between these points if C = 120 for all pipes?

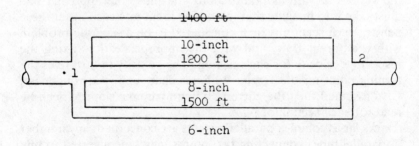

Answer: (H$_L$) = 11 ft

6–4. Water flows from point 1 to point 4 in the network shown below. What equivalent length of 14-in pipe can be substituted for the existing pipes transmitting water between these two points? (C = 100 for all pipes.) *Hints:* (1) Assume a Q; (2) obtain equivalent pipes for series sections 1–2–4 and 1–3–4 to give parallel system; (3) obtain equivalent length of 14-in pipe for the parallel system.

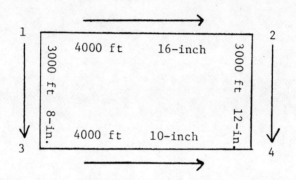

Answer: L = 5000 ft of 14-in pipe

6–5. In the compound pipe system shown in the sketch, the velocity of flow in pipe DC is 3 fps. Using a value of C = 120 for all pipes, compute the required difference in elevation, Z, of the water surface in the two reservoirs, ignoring minor losses.

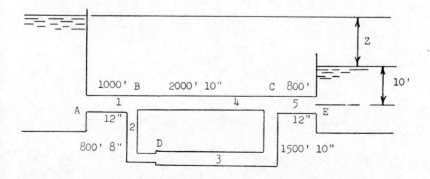

Answer: Z = 31 ft

7

Minor Losses in Fluid Flow

In this chapter, the so-called minor losses (those associated with valves, fittings, etc.) will be considered as a special case. The word *minor* is to merely indicate that the energy loss at these various fittings and valves is small compared to the friction losses that occur in the pipe itself under ordinary circumstances. With very short lengths of pipe, such as in a pipe gallery, the so-called minor losses may become much more important than friction loss in the pipe and, therefore, must be considered. In contrast, in long pipe runs, minor losses are usually neglected since the pipe friction is so large compared to the "minor" losses. In every case, the engineer considers the two categories of loss and then makes his own decision as to whether one or the other can be neglected or whether both must be included.

To facilitate computation of the magnitude of minor losses, a nomograph (prepared by the Crane Company) of the various fittings in common use is included (Figure 7–1). The nomograph presented is based on "the equivalent pipe technique." Any pipe fitting or change in cross section will create some turbulence and, therefore, increase the friction of water particles rubbing against one another. This causes a loss in energy or head due to internal friction within the water mass. Proportionally, there is less energy loss in larger fittings than in small.

The "equivalent pipe technique" merely is a method of estimating how much straight pipe would have created the same amount of energy loss as the fitting in question. There are other approaches to the solution of problems involving minor losses.

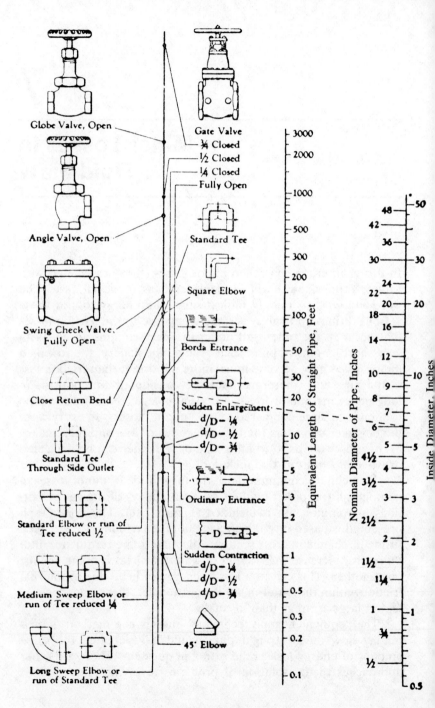

Figure 7–1. Resistance of valves and fittings to flow of fluids.

However, the "equivalent pipe technique" is simple and straight-forward.

The nomograph shown in Figure 7–1 is set up with a great many of the different fittings plotted against the nominal diameter of the pipe or fitting. *Nominal diameter* means the identifying diameter of the pipe. This is the diameter by which pipe is ordinarily described or ordered. The actual inside diameter of various classes and sizes of pipe may be slightly different from the nominal diameter. The actual inside diameter must be either measured directly or taken from a pipe manufacturer's catalog for the pipe in question. The nomograph can be used to determine the equivalent length of pipe by following a straight line from the desired fitting to the nominal diameter. The line will cross the middle scale at a point which gives the equivalent length of that size pipe. This equivalent length of straight pipe will present to the water equal resistance to flow as does the fitting that is being evaluated. This is shown in the following example.

Example 7–1

A pipe has a sudden enlargement going from a 6-in diameter to a 24-in diameter. What is the resistance to flow due to this enlargement?

Consulting the nomograph, we find sudden enlargements classified according to d/D (small diameter/large diameter); in our case, 6/24 or 1/4. Using a straight line, connect the point indicated by the fitting to the nominal diameter of the smaller pipe (6 in) and determine the equivalent length of smaller pipe (6 in) which will result in resistance to flow equal to the sudden enlargement. In this example, as shown by the dotted line, this is 16 ft of 6-in pipe. Note that the equivalent length is always in terms of the smaller diameter. In other words, 16 ft of 6-in pipe produces the same amount of head loss as a sudden enlargement of 6 in to 24 in.

Table 7–1 has been included to provide a direct calculation for smaller size water lines. To obtain the equivalent length of pipe, multiply the equivalent length in pipe diameters by the diameter of the pipe being used.

Table 7-1 Representative Equivalent Length in Pipe Diameters (L/D) of Various Valves and Fittings[a]

	Description of Product		Equivalent Length in Pipe Diameters (L/D)
Globe valves	Conventional	With no obstruction in flat, bevel, or plug-type seat	Fully open	340
		With wing or pin guided disc	Fully open	450
	Y-pattern[b]	With stem 60° from run of pipe line	Fully open	175
		With stem 45° from run of pipe line	Fully open	145
Angle valves	Conventional	With no obstruction in flat, bevel, or plug-type seat	Fully open	145
		With wing or pin guided disc	Fully open	200
Gate valves	Conventional wedge disc, double disc, or plug disc		Fully open	13
			Three-fourths open	35
			Half open	160
			One-fourth open	900
	Pulp stock		Fully open	17
			Three-fourths open	50

Category	Type	Description		Condition	K
				Half open	260
				One-fourth open	1200
Check valves	Conduit pipeline			Fully open	3
	Conventional swing			Fully open	135
	Clearway swing		0.5	Fully open	50
	Globe lift or stop		0.5	Fully open	Same as globe
	Angle lift or stop		2.0	Fully open	Same as angle
	In-line ball	2.5 vertical and 0.25 horizontal	2.0	Fully open	150
Foot valves with strainer		With poppet lift-type disc	0.3	Fully open	420
		With leather-hinged disc	0.4	Fully open	75
Butterfly valves		8-in and larger		Fully open	40
Cocks	Straight through	Rectangular plug port area equal to 100% of pipe area			18
	Three-Way	Rectangular plug port area equal to 80% of pipe area		Fully open	44[c]
		Rectangular plug port area equal to 80% of pipe area		Fully open	140[d]
Fittings	Standard elbow	90°			30
		45°			16
	Long radius elbow	90°			20

Table 7-1   *(Continued)*

Description of Product		Equivalent Length in Pipe Diameters (L/D)
Street elbow	90°	50
	45°	26
Square corner elbow		57
Standard tee	With flow through run	20
	With flow through branch	60
Close pattern return bend		50

aReprinted from Tech. Paper 410, courtesy of Crane Co.
bNo obstruction in flat, bevel, or plug-type seat.
cFlow straight through.
dFlow through branch.

USE OF VELOCITY HEAD

It is sometimes convenient to express certain minor losses in terms of velocity head ($V^2/2g$). This is frequently true of the loss at an entrance to a pipe from a reservoir. With an ordinary entrance condition (square edges) the loss is equal to $1/2\ V^2/2g$. When a pipe is discharging with squared edges and submerged in an open tank and the water in the tank has no velocity, the loss at the exit of the pipe is closer to one velocity head ($V^2/2g$). An example of how this is determined is shown in Example 7–2, since it will give another opportunity where Bernoulli's equation can be applied.

Example 7–2

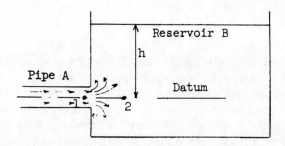

Pipe A discharges into reservoir B creating turbulence and causing an energy loss. A point just inside the pipe is called 1 and a point well into the reservoir where all velocity has ceased is called 2. If a datum through the points is chosen, then Bernoulli's equation reveals the following:

$$\frac{p_1}{w} + \frac{V_1^2}{2g} + Z_1 = \frac{p_2}{w} + \frac{V_2^2}{2g} + Z_2 + (H_L)_p + (H_L)_{Exit}$$

$$\frac{p_1}{w} = h$$

$$\frac{p_2}{w} = h$$

$$Z_1 = 0$$

$$Z_2 = 0$$

$$\frac{V_2^2}{2g} = 0$$

If point 1 is close to the tank so there is no pipe friction, the $(H_L)_p = 0$

$$\cancel{h} + \frac{V_1^2}{2g} + 0 = \cancel{h} + 0 + 0 + 0 + (H_L)_{Exit}$$

Since the h's cancel,

$$(H_L)_{Exit} = \frac{V_1^2}{2g}$$

With an ordinary square-edged connection to a tank or reservoir, the above equation is reasonable. It would not be true for a flared or bell end pipe connection, however, since the entrance velocity is reduced. In these cases, the degree of turbulence is reduced and obviously the energy loss must decrease from that experienced in the case of the square edge connection. Minor losses for pipe configuration other than square-edged connections have been determined experimentally and are available in most hydraulic handbooks.

COMPARISON OF EQUIVALENT PIPE AND VELOCITY HEAD TECHNIQUES

Since the nomograph (Figure 7–1) provides for an ordinary entrance determination, it might be well to check the minor loss at an entrance using the nomograph and compare it with the results of a velocity head calculation.

Example 7–3

As an illustration, let us consider a tank discharging water through a 6-in pipe (ordinary entrance), with a velocity of 3 ft per second.

By the velocity head technique:

$$(H_L)_{ent} = \frac{1}{2}\frac{V^2}{2g} = \frac{1}{2}\frac{(3)^2}{64.4} = 0.07 \text{ ft}$$

By the equivalent technique:

Using Figure 7–1, equivalent pipe = 9 ft.

Using Figure 5–2, 9 ft of 6-in pipe at 3 fps, C = 100 gives a head loss of 9 ft/1000 ft.

$$(H_L)_{ent} = \frac{9}{1000} \times 9 = 0.08 \text{ ft}$$

The student should note that there is not exact agreement here. The value 0.07 is merely close to 0.08 and both are quite acceptable. Precise answers in this area are not possible except where each component is set up and tested separately. Small differences in surface roughness, slight rounding of the edges of the connection, different temperatures, and resulting viscosities of the fluid would produce differences in the resulting energy loss.

Since these variations exist, problems in minor losses must be approached by approximating values and not attempting to get strictly precise answers.

As can be noted from the nomograph, head loss at an entrance is a special case of a sudden contraction. A sudden contraction of d/D = 1/4 is actually very close to a square-edged ordinary entrance. All values of d/D less than 1/4 must lie between the value given for d/D = 1/4 and the value given for an ordinary entrance on the nomograph.

No value is shown on the chart for an ordinary square-edged exit value. However, Example 7–2 has established the fact that loss at exit is $V^2/2g$. This is the same as the Borda entrance as shown below. Whenever a pipe extends into a reservoir or tank a Borda entrance is created.

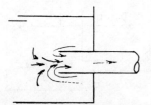

Borda entrance

$$\text{loss} = 1\frac{V^2}{2g} \text{ (approximately)}$$

maximum bending of
flowline = 180°

APPLICATION OF MINOR
LOSS VALUES

With few exceptions, minor losses are determined for the primary purpose of completing a solution using Bernoulli's equation. As long as losses of energy are expressed in feet, values can be inserted directly into the equation and direct solutions made. When losses are expressed in terms of velocity head, insertion into Bernoulli's equation also is possible.

The labor of totaling all the minor losses in a complex system can be enormous. Whether it is decided to neglect this factor or to include it, a valid decision cannot be made in the absence of facts. In other words, when the decision is made to neglect losses, it should be done only after a rough check is made showing their magnitude is small enough to neglect (say, <5%).

Typical problems that arise are somewhat more complex cases of those situations discussed in the two previous chapters. When two values of the three, Q, A, or V, are given, the solutions are straightforward. All fittings are evaluated in terms of equivalent feet of pipe and compared to the actual pipe in the problem. If a significant part of the total is represented by the minor losses, they must be included. They are merely added to the actual length of pipe L to form a somewhat longer theoretical pipe. In all subsequent computations the new length is used in place of the actual length.

STRAIGHTFORWARD SOLUTION

Example 7–4

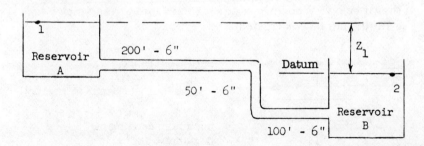

For this example, an identical problem to the one in Chapter 5 where minor losses were neglected will be used. In this way, a basis of comparison can be established. Reference is therefore made to Example 5–7. The sketch and computation previously used can be used again.

As the problem was given, calculate the total head required to keep 1 cfs at steady state.

Setting up Bernoulli's equation:

$$\frac{p_1}{w} + \frac{V_1^2}{2g} + Z_1 = \frac{p_2}{w} + \frac{V_2^2}{2g} + Z_2 + (H_L)_p + (H_L)_{ML}$$

$$0 + 0 + Z_1 = 0 + 0 + 0 + (H_L)_p + (H_L)_{ML}$$

Obviously the next step is to add the pipe losses and the minor losses. Pipe losses = 0.024×350 ft = 8.4 ft (as shown in Example 5–7). Minor losses:

1 entrance	= 9 ft–6 in pipe
2 elbows	= 33 ft–6 in pipe
1 exit (same as Borda entrance)	= 19 ft–6 in pipe

$$\text{Equivalent pipe} = 61 \text{ ft–6 in pipe}$$

$$\text{Minor losses} = 0.024 \times 61 \text{ ft} = \underline{1.5 \text{ ft}}$$

$$\text{Total head required} = 9.9 \text{ ft}$$

Comparison of the results between Example 5–7 and Example 7–4 shows that the neglect of minor losses in this case would create an error of approximately 18% in the total head. This would be entirely too large an error to accept in the usual situation and, hence, inclusion of minor losses must be considered in this case.

SOLUTION OF A MORE COMPLEX SYSTEM

The situation involving a complex piping system with a given total head and the question, "how much water can be delivered under these conditions?" is frequently met. There are other levels of difficulty between this problem and the simpler variety illustrated as Example 7–4.

In Chapter 6, Example 6–2, the problem of an involved piping system was presented with minor losses neglected. The student is advised to again look that problem over and then note the difference in the solution that follows where minor losses are included.

Example 7–5

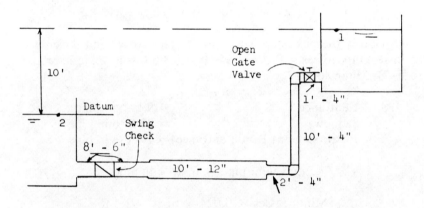

$$\frac{p_1}{w} + \frac{V_1^2}{2g} + Z_1 = \frac{p_2}{w} + \frac{V_2^2}{2g} + Z_2 + (H_L)_{pipe} + (H_L)_{ML}$$

$$0 + 0 + 10 = 0 + 0 + 0 + (H_L)_{pipe} + (H_L)_{ML}$$

or,

$$\text{Total } H_L = (H_L)_{pipe} + (H_L)_{minor\ losses} = 10 \text{ ft}$$

Minor Losses Itemized

$$1 \text{ entrance} = 6 \text{ ft–4 in pipe}$$

$$1 \text{ exit } \frac{V^2}{2g} \text{ (sudden enlargement)} = 19 \text{ ft–6 in pipe}$$

2–4 in elbows at 11 ft each = 22 ft–4 in pipe

1–4/12 enlargement = 8 ft–4 in pipe

1–gate = 2 ft–4 in pipe

1–6/12 contraction = 6 ft–6 in pipe

1–swing check 6 in = 38 ft–6 in pipe

Total minor losses in terms of equivalent pipe = 63 ft–6 in and 38 ft–4 in.

**To Compute
Equivalent Pipes**

Assume 1 cfs flowing

Pipe	H_L ft/1000	ft/ft	Actual loss $Q = 1$ cfs
13 ft–4 in actual	200.0	0.20	2.60 pipe
38 ft–4 in equivalent pipe joint fittings, etc.			7.22 minor loss
8 ft–6 in actual	24.0	0.024	0.2 pipe
60 ft–6 in minor losses			1.5 minor loss
10 ft–12 in actual	0.9	0.0009	0.009 pipe

Total = 11.529
Say = 11.5 ft

To proceed to the selection of an equivalent pipe, any pipe may be chosen, but assuming a 6-in pipe is used, S = 0.024 ft/ft.

For 1 cfs, $\dfrac{11.5}{0.024}$ = 479 ft of 6-in pipe,

which means that 479 ft of straight 6-in pipe gives the same amount of head loss as the entire system shown in the sketch. Substituting the theoretical for the actual system, the actual hydraulic slope is:

$$S = \left(\frac{H_L}{L}\right)$$

$$S = \frac{10}{479} = 0.0209 \text{ ft/ft or } 20.9 \text{ ft/1000 ft}$$

From Figure 5–2 and at this hydraulic slope, only 0.85 cfs could possibly be flowing.

It should be noted that again in this case the problem has accentuated short pipe lengths so as to emphasize the effect of minor losses. The change in flow from Example 6–2 to Example 7–5 indicates 1.85 cfs is reduced to 0.85 cfs when minor losses are considered. This is a substantial change and obviously minor losses could not be neglected for any degree of accuracy.

PROBLEMS

7–1.   What is the loss of head through a 10-in gate valve, 3/4 open?

   Answer: 34 ft of 10-in pipe

7–2.   What is the loss at the entrance to an 8-in pipe from a ground level storage reservoir?

   Answer: 12 ft of 8-in pipe

7–3.   What is the loss through a 12-in gate valve fully open?

   Answer: 6.8 ft of 12-in pipe

7–4.   If a 12-in pipe were suddenly contracted to a 9-in pipe, what would be the loss?

   Answer: 5 ft of 9-in pipe

7–5. Find the equivalent length of pipe for each of the following:

   a. 12-in gate valve 1/2 open
   b. 10-in swing check valve fully open
   c. 12-in–45° elbow
   d. a 12-in pipe suddenly enlarged to a 16-in pipe

   Answer: a.   200 ft of 12-in pipe
         b.   65 ft of 10-in pipe
         c.   15 ft of 12-in pipe
         d.   6.5 ft of 12-in pipe

7–6. Compute the feet of head loss through each of the equivalent pipes in Problems 7–1 through 7–4 using the nomograph (Figure 5–2) and assuming a flow of 500 gpm, and C = 100.

   Answer: 1.   0.095 ft
         2.   0.097 ft
         3.   0.008 ft
         4.   0.023 ft

7–7. Determine the minimum difference in elevation required to maintain a steady flow of 100 gpm through the system shown, C = 100.

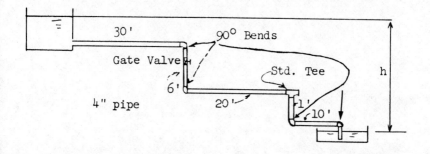

   Answer: h = 1.88 ft

8

Flow in Open Channels

Open channel flow has been defined in Chapter 3 as flow which occurs in a conduit that is not closed or in a closed conduit that is not filled by the flowing water. This type of flow occurs in washwater troughs, settling basins, and usually in sewer lines.

If the flow passing a given cross section does not change with time, the flow is said to be *steady*. Thus, the flow out of an open channel must equal the flow into the channel for the condition of steady flow to prevail. If the cross sectional area and the velocity of the flow remain constant over the entire length of the channel, the flow is further referred to as *uniform*. This lesson will be concerned only with cases where the flow is steady and uniform. Unsteady and nonuniform flow conditions are beyond the scope of this text; for further information on this subject, reference should be made to an appropriate textbook on fluid mechanics or hydraulics.

EQUATIONS GOVERNING FLOW

Many equations are shown in references, texts, and manufacturer's catalogs which state that they are open channel equations, yet they all look somewhat different from each other. All are based on Bernoulli's equation coupled with the concept of a system of forces wherein the weight of fluid sliding down a plane is just opposed by the frictional resistance of the plane on the fluid.

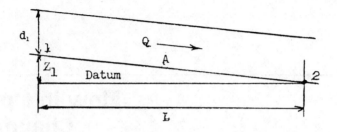

Figure 8–1.

Assume that steady state prevails and that water is flowing down plane A from 1 to 2.

$$\frac{p_1}{w} + \frac{V_1^2}{2g} + Z_1 = \frac{p_2}{w} + \frac{V_2^2}{2g} + Z_2 + (H_L)_{(1-2)}$$

$$\frac{p_1}{w} = \frac{p_2}{w}$$

$$\frac{V_1^2}{2g} = \frac{V_2^2}{2g}$$

$$Z_2 = 0$$

Therefore,

$$(H_L)_{(1\ to\ 2)} = Z_1 \qquad (8.1)$$

This is the same relationship that was shown to exist in the closed pipe where Z_1 was replaced by the differential pressure. In other words, the energy for flow in open channels usually comes from the slope of the channel since the conduit is not pressurized.

In water or sewage plant construction, short lengths of channel are often built with flat bottoms for ease and economy of construction. This does not interfere with flow since the water merely backs up at the upper end of the channel and the surface assumes the proper slope, s, required to discharge the water at the appropriate rate. In other words, the water literally forms its own hill down which to slide.

Equation 8.1 established the fact that a slope (Z_1/L) was necessary to maintain a flow of Q cfs flowing down plane A. This quantity of flow at point 1 has energy due to $V_1^2/2g$, due to p_1/w, and due to elevation Z_1 above the datum plane. Usually the pressure head at point 2, p_2/w, equals the pressure at point 1 (p_1/w) since both are normally atmospheric.

As the fluid moves along, the amount of energy required to keep it moving is just the slope(s) of the channel. The quantity of flow or the velocity head is therefore a function of slope. This is expressed mathematically as:

$$\frac{V^2}{2g} = kS \qquad (8.2)$$

Obviously the energy lost is due to friction with plane A, since the only force opposing motion down the plane is frictional resistance. It can be concluded, therefore, that to maintain a given quantity of flow while the frictional losses were increased would require a steeper slope. Apparently k in Equation 8.2 must involve friction on the wetted surface. Then, k = CR (where R is the hydraulic radius).

Chapter 1 discussed the use of hydraulic radius (R). Chapter 5 also referred to this parameter in the development of the Hazen-Williams equation. From what has been said, it follows that a logical expression for flow in open channels would be:

$$V = C R^a S^b \qquad (8.3)$$

where C = friction coefficient
R = hydraulic radius
S = slope of hydraulic gradient, ft/ft

By experiment and observation researchers have established empirical values for C, a, and b.

$$\text{Chezy: } V = C R^{0.5} S^{0.5} \qquad (8.4)$$

$$\text{Hazen and Williams: } V = 1.318 C R^{0.63} S^{0.54} \qquad (8.5)$$

$$\text{Mannings: } V = \frac{1.486}{n} R^{0.66} S^{0.5} \qquad (8.6)$$

Obviously in the experimental development there has been some difference in the results produced, but all these expressions are close and the precision obtained is probably as good as the ability to select a proper roughness coefficient.

Probably the most popular open channel flow equation today is Mannings', and two monographs are supplied with this chapter to aid in solution of problems using this expression.

CROSS SECTIONS

The geometric properties of cross sections needed for computations in open channel flow are:

1.　the cross sectional area of *flow*, A;
2.　the wetted perimeter, P (the length of the perimeter of the cross section in contact with the water); and
3.　the hydraulic radius, R (the cross sectional area of flow divided by the wetted perimeter).

Typical channels of different cross sectional shapes are illustrated in Figure 8–2. General formulas for computing the geometric properties of the cross sections shown in Figure 8–2 are given in Table 8–1.

CIRCULAR CHANNELS

The circular cross section is certainly the most frequently used open channel. The initial complication which must be faced is that the nomographs provided always yield solutions to problems dealing with full pipes. As long as this restriction is kept in mind, Figure 8–3 will provide a direct solution for roughness of n = 0.013 or n = 0.015. It should be noted that these values cover both cement and vitrified clay pipes as well as coated and uncoated iron (Table 8–2).

A typical example is worked directly on Figure 8–3 so that its use is straightforward and solutions direct.

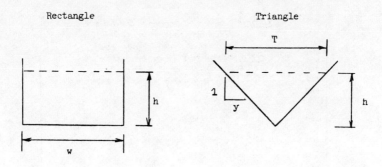

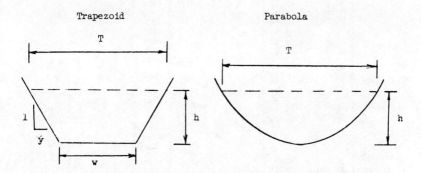

Figure 8–2. Geometric properties of channels.

Table 8–1   Parameters of Various Channels

Cross Section	Rectangle	Triangle	Trapezoid	Parabola
Area, A	wh	yh^2	$wh + yh^2$	$\dfrac{2}{3} hT$
Wetted perimeter, P	$w + 2h$	$2h\sqrt{y^2 + 1}$	$2 + 2h\sqrt{y^2 + 1}$	$T + \dfrac{8h^2}{3T}$
Hydraulic radius, R	$\dfrac{wh}{w + 2h}$	$\dfrac{yh}{2\sqrt{y^2 + 1}}$	$\dfrac{wh + yh^2}{w + 2h\sqrt{y^2 + 1}}$	$\dfrac{2hT^2}{3T^2 + 8h^2}$
Top width, T	w	$2yh$	$w + 2yh$	$\dfrac{3A}{2h}$

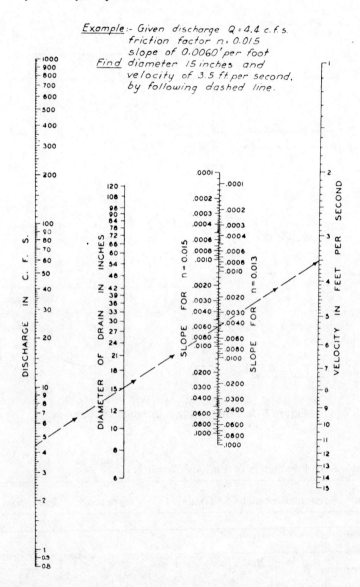

Figure 8–3. Nomograph for computing required size of circular drain, flowing full (n = 0.013 or 0.015).

Table 8–2 Mannings' Roughness Coefficient (n)

Type of Conduit	n
Pipe	
Cast iron, coated	0.012–0.014
Cast iron, uncoated	0.013–0.015
Wrought iron, galvanized	0.015–0.017
Wrought iron, black	0.012–0.015
Steel, riveted and spiral	0.015–0.017
Corrugated	0.021–0.026
Wood stave	0.012–0.012
Cement surface	0.010–0.013
Concrete	0.012–0.017
Vitrified	0.013–0.015
Clay, drainage tile	0.012–0.014
Lined Channels	
Metal, smooth semi-circular	0.011–0.015
Metal, corrugated	0.023–0.025
Wood, planed	0.010–0.015
Wood, unplaned	0.011–0.015
Cement lined	0.010–0.013
Concrete	0.014–0.016
Cement rubble	0.016–0.030
Grass	≈0.2
Unlined Channels	
Earth: straight and uniform	0.017–0.025
dredged	0.025–0.033
winding	0.023–0.030
stony	0.025–0.040
Rock: smooth and uniform	0.025–0.035
jagged and irregular	0.035–0.045

CIRCULAR SECTIONS PARTIALLY FULL

Problems dealing with pipes flowing partially full require a knowledge of trigonometry to obtain the areas and wetted perimeters involved. The more complicated mathematics can be avoided by using a hydraulic elements curve as shown in Figure 8–4. This diagram provides the relationship between the partially filled sec-

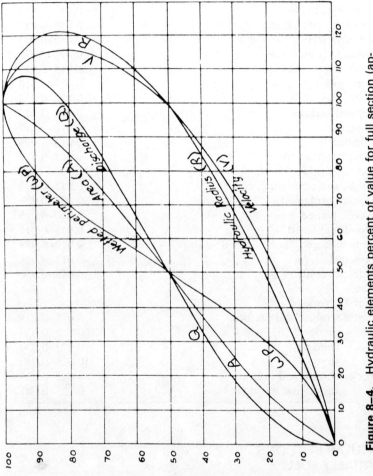

Figure 8-4. Hydraulic elements percent of value for full section (approximate) for circular conduits.

tion and the full section for each of the elements of depth of flow, velocity, area, hydraulic radius, and discharge.

The ratio of depth of flow in the partially filled section h to the depth of flow in a full section (diameter d) is shown on the vertical axis, while the ratio of the other elements for such a partially filled section in terms of the full section are shown by the respective curves and are read on the horizontal axis.

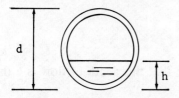

```
depth filled = h
depth full   = d
ratio filled
    to full  = h/d
```

These elements are:

A = area of flow of partially filled conduit

R = hydraulic radius of partially filled conduit

V = velocity of flow in partially filled conduit

Q = quantity of flow (discharge) in partially
     filled conduit

The use of these hydraulic element curves can best be illustrated by means of several examples.

Example 8-1

Consider a 12-in pipe flowing such that the depth of flow is 4 in deep.

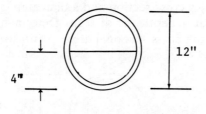

$$h/d = 4/12 = 1/3 = 33\ 1/3\%$$

Using Figure 8–4, the area (A) of flow in the partially filled pipe is 0.28 times the full area of $0.28A_{(full)}$.

The discharge is 0.22 times the discharge from the pipe flowing full or $0.22Q_{(full)}$.

The hydraulic radius R = 0.73 times the hydraulic radius of the pipe flowing full or 0.73 × (d/4).

The velocity V = 0.78 times the velocity through this pipe flowing full.

Example 8–2

Consider a 12-in pipe flowing such that the depth of flow is 6 in (half full).

$$h/d = 0.5$$

Q = 0.5 × the discharge when the pipe is flowing full

A = 0.5 × the area of the pipe flowing full

R = 1 × the hydraulic radius of the pipe flowing full

V = 1 × the velocity in the pipe when flowing full

Note that when a circular pipe flows half full it carries half full discharge with the same velocity as it had when full and has a hydraulic radius equal to that when full.

Example 8–3

A pipe of circular cross section and a diameter, d, of 2 ft carries water flowing at a depth, h, of 6 in. Determine the hydraulic radius, R, and the cross sectional area, A, for the partial flow.

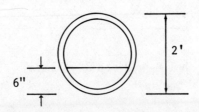

$$\frac{h}{d} = \frac{0.5 \text{ ft}}{2 \text{ ft}} = 0.25$$

From Figure 8–4,

A = 0.2 × the cross sectional area of the pipe

R = 0.6 × the hydraulic radius for the pipe flowing full

$$\text{Cross sectional area of pipe} = \frac{\pi d^2}{4} = \frac{\pi(2)^2}{4} = 3.14 \text{ ft}^2$$

$$\text{Hydraulic radius for pipe flowing full} = \frac{d}{4} = \frac{2}{4} = 0.5 \text{ ft}$$

Therefore, A = 0.2(3.14) = 0.63 ft²

R = 0.6(0.5) = 0.3 ft

The most common problem in sewer work involves a knowledge of procedure between full and partially full sections. The student should now be able to do this type of problem.

Example 8–4

An 18-in sewer has a fall of 2.5 ft between two manholes 250 ft apart. The pipe is flowing at a depth of 7.2 in. What is the quantity of sewage flowing and what would be the velocity at this depth? Also, what is the full capacity of this sewer?

$$S = \frac{2.5}{250} = 0.01 \text{ ft/ft} \qquad \text{Assume n} = 0.015$$

Then, from Figure 8–3,

Full capacity = 9 cfs

Full velocity = 5.2 ft/sec

Then, from Figure 8–4,

$$\frac{h}{d} = \frac{7.2}{18} = 0.4 \text{ or } 40\%$$

At this depth,

Actual discharge = 33% of full value

Actual Q = 0.33 × 9 = 3.00 cfs

Actual velocity = 87% of full value

Actual V = 0.87 × 5.2 = 4.5 fps

SECTIONS OTHER THAN CIRCULAR

Channels in water and sewage plants as well as irrigation ditches all use cross sections which are more difficult to deal with than the more regular circular section. For work in this area, the student must select a roughness from Table 8–2, compute the hydraulic radius, and complete the problem using Figure 8–5, or obtain a direct solution for the velocity using Mannings' formula. Calculators with power functions can be used for a direct solution of the Mannings' equation.

In using this nomograph (Figure 8–5), the coefficient of roughness (Column 1) is lined up with the velocity (Column 5) to obtain a point on the dummy scale (Column 3). This point can then be lined up with the hydraulic radius (Column 2) and extended to Column 4 to obtain the slope. If any three of the four variables are known, the fourth can be obtained. Remember, the two outside scales are used together and the two inside scales are used together with the point on the dummy scale being common.

Example 8–5

An obstruction in a 3-ft-wide rectangular cement-lined channel causes the depth of water to be 2 ft immediately behind it. If the

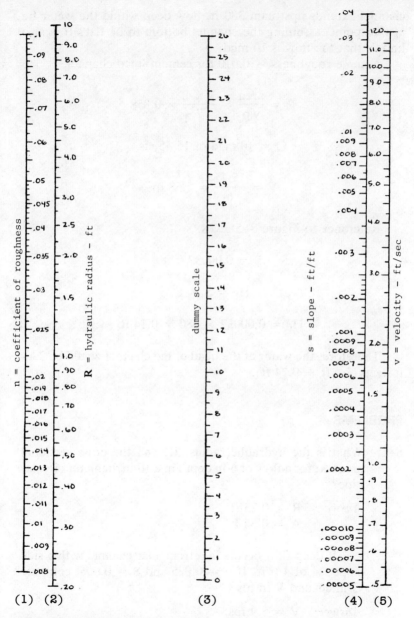

Figure 8–5. Nomograph for Manning equation for all channel shapes; open channel flow.

channel extends upstream 380 ft, how deep would the water be at that point, assuming the channel bottom to be flat? It is also known that the flow is 10 mgd.

Assume roughness = 0.010 for cement-lined channel.

$$R = \frac{\text{area}}{\text{wp}} = \frac{3 \times 2}{7} = 0.858$$

$$Q = 10 \text{ mgd or } 15.45 \text{ cfs}$$

$$V = \frac{15.45}{6} = 2.58 \text{ fps}$$

Reference to Figure 8–5 gives,

$$S = 0.00037 \text{ ft/ft}$$

$$H_L = S \times L$$

$$H_L = 0.00037 \times 380 = 0.14 \text{ ft}$$

Therefore, the water at the head of the channel would be 2.14 ft deep (2.0 ft + 0.14 ft).

PROBLEMS

8–1.   What is the hydraulic radius, R, and the cross sectional area, A, for a flow of 6-in depth in a 10-in diameter circular pipe?

Answer: R = 0.23 ft
           A = 0.34 ft²

8–2.   A 3-ft deep flow occurs in a triangular channel with a side slope, y, of 4 ft/ft. If n = 0.025 and S = 0.006, compute Q in cfs and V in fps.

Answer: V = 5.9 fps
           Q = 213 cfs

8-3. A 20-ft wide rectangular channel lined with corrugated metal carries a flow with a depth of 9 ft. Calculate the rate of flow and the velocity in the channel if S = 0.0004.

Answer: V = 3.45 fps
Q = 621 cfs

8-4. A flow of 400 cfs in a parabolic channel with a roughness coefficient of n = 0.025 has a top width, T, of 30 ft. If S = 0.005, compute the depth of flow, h, and the velocity, V. *Hint:* Use Equation 8.6 (substituting Q/A for V). Compute values of $AR^{2/3}$ for different values of h. Construct a plot of $AR^{2/3}$ vs h. Compute $AR^{2/3}$ for the conditions above, and read corresponding value of h from your plot.

Answer: h = 3.0 ft
V = 6.6 ft

8-5. Determine the minimum hydraulic slope, s, for a rectangular channel that is to carry 600 cfs with a mean velocity of 3 fps. The lining is smooth concrete (i.e., n = 0.014). Assume the channel has a width twice the depth.

Answer: S = 0.000093 ft/ft

8-6. How deep will water flow in a 14-ft rectangular channel that is carrying 615 cfs if S = 0.00075 ft/ft and the channel is lined with smooth concrete? (Use n = 0.014.)

Answer: D = 6.6 ft

9

Flow Measurements
I

Flow measurement is of extreme importance to water and sewage works personnel, since the quantities of product produced or treated must be evaluated. Flow measurements are also important in determining the distribution in various portions of a water system, and for controlling and evaluating the quantities of sewage discharge to our waterways. The flow expression of most interest is rate of flow, represented by Q in the general flow relationship, Q = AV which has been previously presented and explained (Equation 3.1). Common units for expressing the rate of flow are cubic feet per second (cfs), gallons per minute (gpm), and million gallons per day (mgd).

In considering closed pipe flow under pressure, the rate of flow can be evaluated by placing a constriction within the pipe. This constriction produces a sudden decrease in area, and a resulting increase in velocity and velocity head. The total head is made up of $(p/w) + (V^2/2g)$ (assuming elevation constant). Since the velocity head is increased, p/w must decrease. As a result, it must be apparent that changes in p/w are related to changes in velocity. Since quantity is related to velocity through Q = AV, pressure head changes can be correlated with flow.

Flow meters which reduce the cross sectional area and which are commonly used in water and sewage works practice are orifices and venturi meters. Such meters have no moving parts and act merely to create a pressure differential across the constriction. This differential may read in feet of pressure and be converted to units of flowrate.

Many people think of the common type of water meter used in measuring water to homes and businesses as the only kind of water meter in existence. Actually, the household water meter is of little value in measuring large quantities of flow. However, such a unit is uniquely suited to measuring residential use because it is a positive displacement type of device. Each time a unit of water passes through the meter, a disc oscillates. This in turn operates a gear train to record amounts of flow. A meter of this type for large quantities of flow would be exceedingly expensive and really unnecessary. The propeller-type meter, the orifice, or the venturi meter would be the device used in such a case.

ORIFICES

The orifice, when used as a flow metering device, consists of a circular hole in a thin metal plate which is fixed at the discharge point of a tank or reservoir, or clamped between the flanges of a pipe. These two uses of an orifice are illustrated in Figure 9–1.

For the orifice application indicated in Figure 9–1a, the rate of flow from the reservoir can be evaluated through use of Bernoulli's equation and the equation of continuity. The velocity of the water discharging through the orifice can be determined by writing Bernoulli's equation between points 1 and 2 as follows:

$$\frac{p_1}{w} + \frac{V_1^2}{2g} + Z_1 = \frac{p_2}{w} + \frac{V_2^2}{2g} + Z_2 + (H_L)_{(1-2)}$$

If the datum is placed through point 2, Z_2 can be eliminated from the equation. Also, since there is a free discharge of water

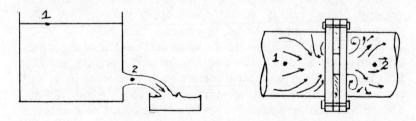

Figure 9–1. Two uses of an orifice.

to the atmosphere, the pressure heads at points 1 and 2 are zero
and can be eliminated, thus leaving the equation in the following
form (when head loss is neglected):

$$Z_1 + \frac{V_1^2}{2g} = \frac{V_2^2}{2g}$$

It is also apparent that the cross sectional area of the reservoir
is large in comparison to the orifice area and the downward velocity is virtually zero. Therefore, the velocity head at point 1
becomes negligible and can be eliminated. This leaves:

$$\frac{V_2^2}{2g} = Z_1 \qquad (9.1)$$

Then, solving for the velocity at point 2:

$$V_2 = \sqrt{2g\,Z_1}$$

where Z_1 = the elevation head producing flow through the orifice.

By knowing the cross sectional area (A) of the opening and
using the equation of continuity, $Q = AV$, the rate of flow through
the orifice can be expressed as:

$$Q = A\sqrt{2g\,Z_1} \qquad (9.2)$$

With Z_1 expressed in feet and A in square feet, Q is evaluated
in terms of cubic feet per second (cfs). Equation 9.2 gives the
theoretical rate of flow, since loss of head through the orifice was
neglected. To evaluate correctly the actual rate of flow, the head
loss through the orifice must be considered. This loss is usually
taken into account by applying an experimental orifice discharge
coefficient (C_d) in the following manner:

$$Q = C_d\,A\sqrt{2g\,Z_1} \qquad (9.3)$$

The orifice discharge coefficient depends on the sharpness of
the edge of the orifice. Generally accepted coefficient values range
from 0.60 for a sharp-edged orifice to 0.98 for round-edged orifices.

Example 9–1

A 4-in diameter sharp-edged orifice discharges water from a reservoir under an elevation head Z_1 of 20 ft. What is the rate of flow through the orifice?

$$Q = C_d A \sqrt{64.4 \times Z_1}$$

Assume: $C_d = 0.60$

$$A = \frac{\pi d^2}{4} = \frac{\pi}{4}\left(\frac{4}{12}\right)^2 = 0.087 \text{ ft}^2$$

$$Q = 0.60 \times 0.087 \sqrt{64.4 \times 20}$$

$$Q = 0.60 \times 0.087 \sqrt{1288}$$

$$Q = 0.60 \times 0.087 \times 35.9$$

$$Q = 1.87 \text{ cfs}$$

PIPE FLOW

The use of an orifice to measure the rate of flow in a pipe under pressure is usually similar to that illustrated in Figure 9–1b.

$$\frac{p_1}{w} + \frac{V_1^2}{2g} + Z_1 = \frac{p_2}{w} + \frac{V_2^2}{2g} + Z_2 + (H_L)_{(1-2)}$$

Neglecting head loss and $V_1^2/2g$, and with $Z_1 = Z_2$, the equation becomes:

$$\frac{V_2^2}{2g} = \frac{p_1}{w} - \frac{p_2}{w}$$

$$V_2 = \sqrt{2g\,(p_1 - p_2)/w} \qquad (9.4)$$

Equation 9.4 neglects $V_1^2/2g$, which is only of some magnitude when the approach pipe and the orifice diameter are relatively close in size. The $V_1^2/2g$ term adds to the total head, and thus, if neglected, causes the flow to be underestimated.

If more accuracy is desired, Equation 9.8 (developed later in this chapter) may be used.

Example 9–2

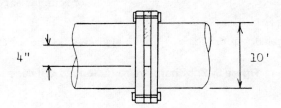

In the sketch as shown, assume the orifice has a diameter of 4 in, a coefficient of discharge of 0.65, and is inserted in a 10-in flanged pipe. If the differential pressure across the orifice plate is 21 ft of water, what is the flow?

Neglecting the approach velocity head:

$$Q = C_d A \sqrt{2g \frac{(p_1 - p_2)}{w}}$$

$$Q = 0.65 \times 0.087 \times \sqrt{64.4 \times 21}$$

$$Q = 2.08 \text{ cfs} \left(\text{neglecting } \frac{V_1^2}{2g}\right)$$

It should be noted that Q is not an exact answer. Even if the more accurate equation (9.8) is used, the answer is still not exact. The reason is because of a lack of knowledge of the exact value of the coefficient of discharge. The so-called sharp-edged orifice is cut with a 90° edge as is shown in Figure 9–2a.

Such an orifice has a definite coefficient of discharge. With use, the edge becomes rounded and the coefficient increases. Ori-

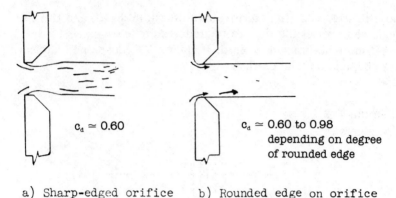

c_d ≈ 0.60   c_d ≈ 0.60 to 0.98
depending on degree
of rounded edge

a) Sharp-edged orifice   b) Rounded edge on orifice

Figure 9–2.   Sharp- vs round-edged orifice.

fice plates therefore lose accuracy with age and such a limitation
should be recognized.

A further limitation on the use of the orifice plate is the extreme turbulence created and the resulting energy loss. Unfor-
tunately, this is not the only loss involved in the installation. As
the water moves through the orifice, a very high velocity results
which is discharged into a body of water having a relatively low
velocity. In effect, most of the energy of the jet is dissipated in
turbulence in the pipe that follows.

The amount of the loss due to turbulence varies depending
on the relative size of the orifice and the pipeline into which it is
inserted. If the orifice is small and it discharges into a fairly large
pipe, the loss approaches $V_2^2/2g$, similar to the loss at the exit of
a pipe as previously defined (Example 7–2). If the orifice is large
in contrast to the pipe, the water downstream of the orifice in the
pipe has appreciable velocity and as a net result, less energy is
lost. The total energy losses through an orifice plate are significant. These losses must be evaluated in terms of "how much does
it cost in terms of power?" Savings in electrical energy may pay
for the installation of a more expensive meter which will have
better flow characteristics.

The advantages of orifice plates are their extreme simplicity
and, hence, economic first cost. Perhaps even more importantly
is the small space required for such a meter.

The only precaution necessary is that approximately 10 diameters of straight pipe should precede the orifice plate to create a smooth pattern of flow to the meter. In addition, about five diameters of straight pipe following the meter is desirable. If these minimum conditions are not met, the meter will probably discharge at a somewhat lower capacity than expected. Standard gauge connections are also important. Usually the locations of the gauge connections are specified for commercial orifice plates. Otherwise, the orifice plate must be calibrated on the site.

VENTURI METER

The other form of reduced area metering devices in common use is the venturi meter as shown in Figure 9–3.

This device is normally installed in a straight length of pipe as a unit. It consists of a short nozzle connected to a meter throat, followed by a smooth diffusion exit cone with an angle of flare. As water flows through this meter, the normal hydraulic pressure exists at the entrance, and a reduced pressure is created in the meter throat. Again, as with the pipe orifice, the pressure difference between sections 1 and 2 is transferred to a differential gauge or recorder that usually indicates the rate of flow in cfs, gpm, or mgd. To have continually accurate flow measurements, the small tubes or annular spaces connected to the gauge or recorder must

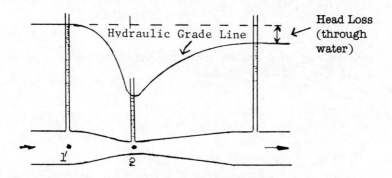

Figure 9–3. Venturi meter.

be kept clean and free of solids as well as air bubbles. This would be especially difficult with sewage where grease is a continual problem.

This does not imply that gauge connections are not a problem with orifices. In fact, the problem is about equivalent in either case.

The great advantages exhibited by venturi sections in contrast to the orifice plates are, first, the wear and consequent loss of accuracy in an orifice is not a problem in the venturi. Second, the loss of energy in turbulence downstream from the meter is not nearly as great with the venturi section. The diffuser cone provides a gentle, yet constant conversion of velocity head to pressure head and thus recovery of a large portion of $V_2^2/2g$ energy is possible.

The disadvantages of the venturi meter are its long length and its initial cost. Both of these factors may outweigh the energy loss and loss of accuracy inherent in the aging of an orifice meter.

Where particulate matter is concerned, as in sewage, the venturi is almost always used since the solids can be scoured through easily, and wear in the throat does not imply too many problems. Solids do not easily pass the orifice plate, and if they do, the sharp edge usually suffers rapid deterioration.

Quantity of discharge through venturi meters can also be calculated using the Bernoulli and continuity equations. Writing the Bernoulli equation between points 1 and 2 as shown in Figure 9–3 gives:

$$\frac{p_1}{w} + \frac{V_1^2}{2g} + Z_1 = \frac{p_2}{w} + \frac{V_2^2}{2g} + Z_2 + (H_L)_{(1 \text{ to } 2)}$$

Neglecting head loss and assuming the venturi is level ($Z_1 = Z_2$), the above equation becomes:

$$\frac{p_1}{w} + \frac{V_1^2}{2g} = \frac{p_2}{w} + \frac{V_2^2}{2g}$$

Rewriting this equation gives:

$$\frac{p_1 - p_2}{w} = \frac{V_2^2 - V_1^2}{2g} \tag{9.5}$$

From the continuity equation $(A_1V_1 = A_2V_2)$, we know that:

$$D_1^2V_1 = D_2^2V_2 \text{ or } V_1 = \left(\frac{D_2}{D_1}\right)^2 V_2 \text{ and } V_1^2 = \left(\frac{D_2}{D_1}\right)^4 V_2^2$$

Substituting this into Equation 9.5 gives:

$$\frac{p_1 - p_2}{w} = \frac{V_2^2 - \left(\frac{D_2}{D_1}\right)^4 V_2^2}{2g}$$

Rearranging this gives:

$$\frac{p_1 - p_2}{w} = \frac{V_2^2\left[1 - \left(\frac{D_2}{D_1}\right)^4\right]}{2g}$$

and solving for V_2^2 gives:

$$V_2^2 = \frac{2g\,(p_1 - p_2/w)}{[1 - (D_2/D_1)^4]}$$

or:

$$V_2 = \sqrt{\frac{2g\,[(p_1 - p_2)/w]}{[1 - (D_2/D_1)^4]}} \tag{9.6}$$

This is the theoretical velocity through the throat of the venturi section. The actual velocity can be determined by multiplying the theoretical velocity by the coefficient of discharge, C_d. Therefore, the velocity in Equation 9.6 is actually:

$$V_2 = C_d \sqrt{\frac{2g\,[(p_1 - p_2)/w]}{1 - (D_2/D_1)^4}} \tag{9.7}$$

By continuity $Q = V_2A_2$, Equation 9.7 becomes:

$$Q = C_dA_2 \sqrt{\frac{2g\,[(p_1 - p_2)/w]}{1 - (D_2/D_1)^4}} \qquad (9.8)$$

$(p_1 - p_2)/w$ is the difference between the upstream and throat pressures in the venturi in feet of water.

The coefficient of discharge (C_d) for venturi meters is much higher than for orifice plates since there is less turbulence due to the smooth flow transition. The coefficient of discharge for a venturi meter may vary from 0.96 to 0.98 depending on the flare of the exit cone. This equation can also be used for a direct solution of a pipe orifice problem where the upstream velocity is significant, as indicated previously in this chapter.

Example 9–3

Given a venturi section with an upstream diameter of 12 in and a throat diameter of 10.5 in, the discharge coefficient for the venturi meter is 0.96, the pressure difference $[(p_1 - p_2)/w]$ is 2 in of water, calculate the flow.

Using Equation 9.8:

$$Q = C_dA_2 \sqrt{\frac{2g\,[(p_1 - p_2)/w]}{1 - (D_2/D_1)^4}}$$

$$C_d = 0.96$$

$$A_2 = \frac{\pi \times \left(\dfrac{10.5}{12}\right)^2}{4} = 0.601$$

$$\frac{D_2}{D_1} = \frac{10.5}{12} = 0.875$$

$$\left(\frac{D_2}{D_1}\right)^4 = (0.875)^4 = 0.586$$

$$\frac{p_1 - p_2}{w} = \frac{2}{12} = 0.167 \text{ feet of water}$$

$$Q = 0.96 \times 0.601 \sqrt{\frac{64.4 \times 0.167}{1 - 0.586}}$$

$$Q = 2.94 \text{ cfs}$$

There is a trend toward the development of so-called insert flow tubes or nozzles which have a shorter laying length and slip directly inside the regular pipe. The shorter length can only come about by a greater angle of flare in the exit cone. This in turn must reflect in a less efficient conversion of velocity to pressure at this point, causing the total head loss to increase.

RATE OF FLOW CONTROLLER

In water treatment plants, the flow through rapid sand filters is normally controlled by some form of flowrate controller. The operation of some rate controllers is based on the venturi meter principle. A cutaway illustration of such a rate controller is shown in Figure 9–4.

The need for such a unit is generally brought about because of the accumulation of suspended matter in the sand bed during the filter cycle. The filter normally begins its operation with about a foot of head loss. At the termination of the cycle, the loss is perhaps 8 feet of water. Since flow through the bed is a direct function of the head loss, such a sand bed would have very high flows initially in contrast to the final flow. The rate controller is a type of valve which opens up, and thus reduces head loss through the valve, as the head loss in the sand bed accumulates. Thus, the total head loss through the system is kept constant and the flow through the sand bed is likewise constant.

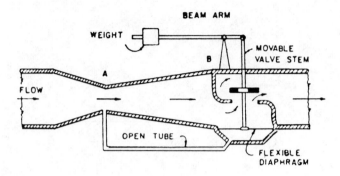

Figure 9–4.   Rate of flow controller.

The valve mechanism is controlled by the flow through a venturi meter. The pressure differential created regulates a flexible diaphragm which adjusts the valve opening in the controller so that a constant rate of flow is maintained.

MANOMETERS

A manometer is an instrument used for measuring pressure. Differential manometers are often used for determining differences in pressure across orifices, venturi meters, and other flow measuring devices. The simplest type of manometer is illustrated in Figure 9–5. Water rises in the tube which is connected directly to the conduit. The pressure intensity is measured by the vertical distance, h, from the liquid surface in the tube to the point where the pressure is to be measured (center of conduit). In this case, the pressure intensity is expressed in terms of feet of the liquid (water), which is rising directly from the conduit into the manometer tube. If the specific gravity of the liquid is S (see Chapter 1 for definition of specific gravity), then the pressure intensity, in feet of water, at point A is:

$$p_A = hS$$

For some cases, the distance, h, may be impractical if the pressure in a pipe is high. In such cases, a second liquid of greater

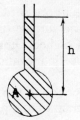

Figure 9–5. Example of a simple manometer.

specific weight, such as mercury, is employed. A differential manometer, the type most commonly used in water and waste treatment works, determines the difference in pressure intensities at
two points, A and B, when the actual pressure at any point in
the system is unknown. An example of a differential manometer
is given in Figure 9–6.

In any differential manometer problem, the general procedure to be followed may be outlined:

1. Start at one end of the manometer and write the pressure
    there, either as a known quantity or as an approximate sym-
    bol if the value is not known.

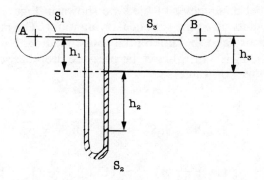

Figure 9–6. Differential manometer.

2. Add to this the change in pressure intensity from one liquid surface to the next liquid surface (plus if the next liquid surface is lower, minus if higher). For feet of water this is the product of the difference in elevation in feet and the specific gravity of the liquid.
3. Continue until the other end of the gauge is reached, and set the resulting expression up to that point equal to the pressure at that point, known or unknown.

It is essential that one use *consistent units* when writing such an expression for manometers.

Applying the procedure as outlined to the example of a differential manometer as shown in Figure 9–6, one obtains:

$$p_A + h_1 S_1 + h_2 S_1 - h_2 S_2 - h_3 S_3 = p_B$$

$$p_A - p_B = - h_1 S_1 - h_2 S_1 + h_2 S_2 + h_3 S_3$$

$$p_A - p_B = - h_1 S_1 + h_2(S_2 - S_1) + h_3 S_3$$

If $h_1 = h_3$ and $S_3 = S_1$, as in a normal manometer, then:

$$p_A - p_B = h_2(S_2 - S_1)$$

Example 9–4

A differential manometer of the type shown in Figure 9–6 connects two points on a pipe system carrying water. The manometer fluid is mercury and the differential gauge reading, h_2, is 5 in. If h_1 is 4 in and h_3 is 4 in, what is the pressure drop between points A and B in feet of water?

Solution:

$$p_A + h_1 S_1 + h_2 S_1 - h_2 S_2 - h_3 S_3 = p_B$$

$$p_A + \frac{4}{12}(1) + \frac{5}{12}(1) - \frac{5}{12}(13.6) - \frac{4}{12}(1) = p_B$$

$$p_A - p_B = \frac{5}{12}(13.6 - 1)$$

$$p_A - p_B = 5.24 \text{ ft of water}$$

PROBLEMS

9–1. What is the discharge through a 200-by-100-in venturi meter for a 2-in reading on a water-air differential manometer if $C_d = 0.99$?

Answer: Q = 183 cfs

9–2. Consider the system shown in the drawing below. Compute: (a) the discharge through the 4-in orifice if $C_d = 0.85$, and; (b) the head loss for the flow computed in part a.

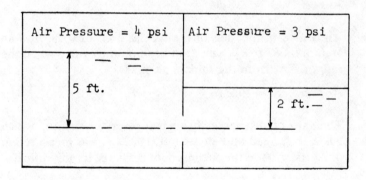

Answer: a. Q = 1.37 cfs
b. H_L = 5.31 ft

9–3. Compute the gauge difference on a mercury-water differential manometer for a 5 cfs discharge through an 8-by-6-in venturi meter for $C_d = 0.983$.

Answer: 0.568 ft

9–4. Determine the flow through a 4-in diameter orifice (C_d = 0.68) installed in a 6-in pipe when the gauge difference is 2 in on a mercury-water differential manometer.

   Answer: Q = 0.688 cfs (neglecting upstream velocity head)
   Q = 0.768 cfs (using the venturi equation)

9–5. A 6-in diameter orifice with C_d = 0.63 is installed in a 12-in diameter water line. What gauge difference would be required on a mercury-water differential manometer for a flow of 11.4 cfs?

   Answer: 10.50 ft (neglecting upstream velocity head)
   9.84 ft (using the venturi equation)

9–6. Neglecting any head losses through a reservoir orifice, what is the rate of flow in cfs from the orifice when there is a 9-ft head on the 4-in diameter orifice?

   Answer: Q = 2.1 cfs

9–7. An orifice in the side of a tank has a C_d of 0.98, and an area of 0.087 ft². What is the head on the orifice when the rate of flow from the tank is 3.0 cfs?

   Answer: h = 19.2 ft

9–8. The rate of discharge through a reservoir orifice, which has a 16-ft head and an area of 0.022 ft², was measured at 0.513 cfs. What is the coefficient of discharge for this orifice?

   Answer: C_d = 0.73

10

Flow Measurements II

The measurement of flow in open channels or in pipes which normally flow partially full, such as sewers, is of importance in water and sewage works practice. Measuring devices which are generally used for this type of flow measurement are weirs or calibrated flumes. In the case of a free discharge from a pipe, it is also possible to measure certain characteristics of the discharge pattern and relate these to the rate of flow.

WEIRS

For the measurement of open channel flow, weirs find wide application. Weirs can be generally thought of as regularly shaped obstructions over which water flows. In the simplest and most usual form, weirs are made of smooth, flat metal plates with sharpened edges at the point of overflow. In this form, a weir is merely an orifice with flow taking place over a restricted portion of its edge. The classification of weirs is in accordance with their shape. The most commonly used weirs are the rectangular weir and the 90° V-notch weir as indicated in Figure 10–1.

THE 90° V-NOTCH WEIR

The V-notch weir is best suited for measuring small flows. In this case, the head for a given flow is accentuated and the accuracy in measurement enhanced. This type of weir can be easily placed

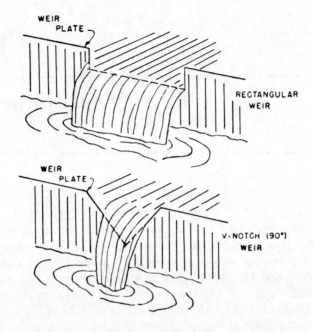

Figure 10–1. Rectangular and V-notch weirs.

in a small channel or in a sewer manhole. The general equation for flow over a 90° V-notch weir is:

$$Q = 2.5 \, h^{2.5} \qquad (10.1)$$

where h = the elevation head on the weir, as shown in Figure 10–2.

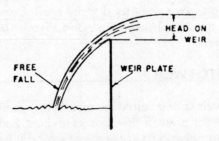

Figure 10–2. Side view of weir showing elevation head.

The rate of flow in cfs is obtained when h is expressed in feet. To arithmetically solve Equation 10.1, $h^{2.5}$ can most easily be thought of as the square root of h^5, or $\sqrt{h \times h \times h \times h \times h}$. The evaluation of Q is simplified through the use of a graph such as Figure 10–3. The equation can be solved directly with a calculator with a power function.

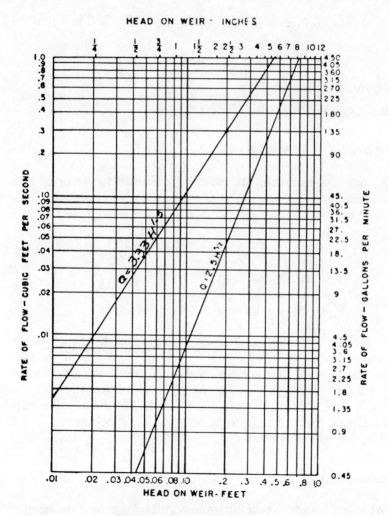

Figure 10–3. V-notch weirs.

Example 10–1

What is the rate of flow over a 90° V-notch weir if the head on the weir is measured as 0.5 ft?

Arithmetic Solution:

$$Q = 2.5 \, h^{2.5} \text{ or } 2.5 \, h^{5/2}$$

$$h^{5/2} = (0.5)^{5/2} = \sqrt{0.5 \times 0.5 \times 0.5 \times 0.5 \times 0.5} = 0.18$$

$$Q = 2.5 \times 0.18 = 0.45 \text{ cfs}$$

Figure 10–3 Solution:

Read Q directly as 0.45 cfs or 202 gpm.

RECTANGULAR WEIRS

A sharp-edged rectangular weir is best suited for accurate measurement of large flows. The flow equation for a standard rectangular weir is:

$$Q = 3.33 \, Lh^{1.5} \qquad (10.2)$$

where Q = rate of flow, cfs
　　　L = width of weir opening, ft
　　　h = head on the weir, ft

In the arithmetic solution of this equation, $h^{1.5}$ or $h^{3/2}$ can be considered as the square root of h^3, or $\sqrt{h \times h \times h}$. A more convenient method of solution of this equation is available through the use of Figure 10–3. This graph provides flow values for L of 1 ft and heads up to 0.45 ft. For other widths the result must be multiplied by the width of the weir, L. For heads above 0.45 ft, the equation must be solved.

Example 10–2

What is the flow over a 2-ft-wide rectangular weir if the head on the weir is measured to be 0.4 ft?

Arithmetic Solution:

$$Q = 3.33 \ (2) \ (0.4)^{3/2}$$

$$(0.4)^{3/2} = \sqrt{0.4 \times 0.4 \times 0.4} = \sqrt{0.064} = 0.253$$

$$Q = 3.33 \times 2 \times 0.253 = 1.68 \text{ cfs or } 756 \text{ gpm}$$

Figure 10–3 Solution:

Read Q directly as 0.84 cfs for 1-ft weir.

$$2 \times 0.84 = 1.68 \text{ cfs or } 756 \text{ gpm}$$

When a weir is used to evaluate the rate of flow in an open channel or partially full pipe, the head on the weir is measured with some form of hook or point gauge, staff gauge, or continuous water surface elevation recorder. To ensure accurate head mea-surements, the zero point of the gauge or recorder chart must be precisely referenced to the overflow edge of the weir. Also, the head on the weir should always be measured far enough upstream from the weir so that the contraction of the water surface does not affect the head readings.

ADVANTAGES AND
DISADVANTAGES OF WEIRS

The major advantage of a weir is the ease with which one can be installed in any system of channels. Practically any tin shop can fabricate a weir using regular galvanized sheet iron. It can be erected with a framework of lumber using plywood for strength and backup, and the sheet iron for the control section.

Often a tank or other appurtenances of a system will have a weir as part of its normal construction to control depth behind it. This would be true of all clarifiers, chlorine detention tanks, etc. Such weirs merely need to be equipped with gauges to provide the necessary metering.

A weir placed in a channel causes the water to back up, and this may create problems upstream from the weir. Such backup in sewers is sometimes critical because of surcharge problems. All sewers are designed to maintain scouring velocities of 1.5 to 2.5

fps. The reduction of velocity due to the partial blockage of a weir causes the sewer to act like a long narrow sedimentation basin. This can be a major problem.

Weir plates in sewers also are affected by the solids such as strings or rags which get caught on the lip of the weir. Constant supervision is necessary if a continuous recording of weir flow is being made.

For a weir to function as described in the equations and curves in this chapter, the nappe or jet created must "leap" free from the sharp edge and fall freely to the channel below. If the nappe clings to the edge, dribbles down the plate or does not create a jet which is completely aerated as it falls, measurements from the weir will probably be inaccurate.

Such a fall may also be a problem. In effect, since it is necessary for accuracy, it must be considered a vital part of the weir. Yet, this difference in water level from a point above the weir to a point in the channel below constitutes loss of head and where conservation of head is vital, a weir may be impossible to use.

FLOW FROM THE END OF A PIPE

If a pipe or sewer has a free discharge into the air, the horizontal and vertical projections of the jet of flowing water can be measured, and these measurements then used to compute the rate of flow from the pipe. This method of flow measurement may be used where an occasional measurement is required and it is too difficult or too expensive to install one of the previously mentioned, more conventional flow measuring devices.

Water will fall in response to gravity just as any solid particle. No matter what the velocity or horizontal travel, the particle will begin to fall as the water leaves the pipe and in a period of one second will be accelerated from a vertical speed of zero initially to g, or 32.2 fps. Velocity is equal to acceleration multiplied by time or $V = gt$. Since the initial velocity is 0, the average velocity is $V = \dfrac{0 + gt}{2}$ or $V = 1/2\ gt$. The distance an object travels equals velocity multiplied by time. Therefore, the vertical distance traveled (y) is $y = Vt$. Substituting for V:

$$y = 1/2 \text{ gt(t) or } y = 1/2 \text{ gt}^2 \qquad (10.3)$$

The horizontal distance of travel is unaffected by this vertical acceleration and the fluid should continue on its way at the same velocity it had when leaving the pipe. If we call the distance traveled horizontally x, its value would be:

$$x = Vt \qquad (10.4)$$

Figure 10–4 indicates the method of measurement. Corresponding central points in the jet should be selected for reference. A large T-square of wood helps in selecting the central points and in making accurate measurements of the x and y distances.

From Equations 10.3 and 10.4, the term t^2 can be eliminated.

$$x^2 = V^2 t^2 \text{ and } t^2 = \frac{x^2}{V^2}$$

Therefore,

$$y = \frac{1}{2} g \frac{x^2}{V^2}$$

or,

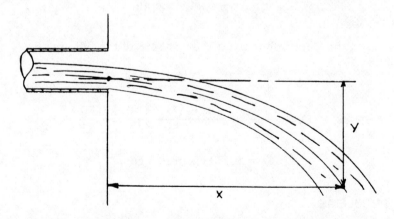

Figure 10–4.   Coordinates of a jet.

$$V = \frac{4\,x}{\sqrt{y}} \qquad (10.5)$$

$Q = AV$ and Equation 10.5 can be altered to give quantity by multiplying both sides of the equation by A.

$$Q = \frac{4\,A\,x}{\sqrt{y}} \qquad (10.6)$$

The value of A can be obtained by measuring the depth of flow at the end of the pipe and computing the ratio of h/d. With this value, the actual area of flow follows from the hydraulic elements curve, Figure 8–4. Knowing the area of the full section, the area of the partially full section can be computed and inserted in Equation 10.6, above.

Example 10–3

An 8-in pipe discharges a stream of water estimated to be 4 in deep at the end of the pipe. Horizontal and vertical distances to the center of the jet are 1 and 1.5 ft, respectively. What quantity of liquid is being discharged?

$$\frac{h}{d} = \frac{4}{8} = 50\% \text{ depth}$$

At this depth, the area of flow is 50% of the full area.

$$A = 0.50 \times 0.349 = 0.174 \text{ ft}^2$$

$$Q = \frac{4\,A\,x}{\sqrt{y}} = \frac{4 \times 0.174 \times 1.0}{(1.5)^{1/2}} = \frac{0.70}{1.22}$$

$$Q = 0.570 \text{ cfs or } 256 \text{ gpm}$$

PROBLEMS

10–1.   A rectangular channel has a maximum discharge of 60 cfs. At this flow, the depth in the channel was measured at 1 ft and the width 80 ft. Assume that a rectangular weir is to be built which will measure the flow without increasing the depth of the water to more than 2 ft. How high should the weir plate be? Assume the length of the weir is 80 ft.

Answer: 1.63 ft

10–2. A triangular 90° V-notch weir is to be built in a manhole to discharge 400 gpm. What head should be provided for?

Answer: 0.66 ft

10–3. The head obtained in Problem 10–2 will probably cause excessive amounts of backup for the system in which it is installed, so it is decided to replace the V-notch with a rectangular weir having no more than 0.2 ft head. The designer is now concerned about the width of the manhole. How wide a crest would be needed?

Answer: 2.99 ft

10–4. An 18-in sewer on a grade of 0.002 is presently flowing at 30% of full depth, and it will be metered using a 90° V-notch weir. Allowing for a 0.2-ft drop to the water surface below the weir, what would be the depth in the sewer above the weir, and what would be the velocity in the sewer at this depth and flow? Assume n = 0.015.

Answer: New depth = 15.1 in
New area of flow = 1.53 ft^2
Velocity = Q/A = 0.722/1.53 = 0.47 fps

10–5. A weir plate for a 90° V-notch weir is to be laid out by a sheet metal fabricator. The range of flow desired is 400–1000 gpm. If a minimum of 2 in of sheet iron is needed for bolting to a plywood backup board, what minimum-sized rectangular sheet of 12-gauge metal would be needed?

Answer: H = 0.955 ft
Size of sheet metal = 18.2 × 18.2 in

11

Pump Types and Characteristics

If two reservoirs are connected by a pipe as shown in Figure 11–1, the flow from point 1 to 2 is determined by the difference in elevation between the two water surfaces. The maximum amount of water the pipe connecting the two reservoirs can carry is limited by the total head in the system and it can be no more than Z_1, the difference in elevation.

For example, if reservoir B is 20 ft lower than A, and 1000 ft of 6-in pipe connects the two reservoirs, what is the rate of flow through the 6-in pipe? By referring to the nomograph in Chapter 5, Figure 5–2, a 6-in pipe with a head loss of 20 ft/1000 ft and a C factor of 100 is found to have a flow of 380 gpm. The velocity in the pipe will be 4.43 fps. Minor losses at entrance, $\frac{1}{2}(V^2/2g) = 0.152$ ft, and at exit, $V^2/2g = 0.3$ ft, are neglected because they are very small in comparison to the total head loss.

If twice as much water, 760 gpm, is desired through the 6-in

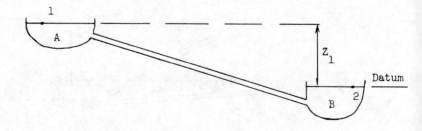

Figure 11–1. Two reservoirs connected by a pipe.

135

line between reservoirs A and B, the head loss due to friction is equal to 74 feet of head (see Figure 5–2). For this to flow by gravity, the elevation of the water surface of reservoir A would have to be raised 54 ft higher (see Figure 11–2) than it is now. With the reservoirs fixed at a differential head of 20 ft, the only way to increase the flow would be to introduce a pump, which will provide 54 ft of added head and, in doing so, cause the flow to increase to 760 gpm from reservoir A to B. The additional head required is due to an increase in friction in the pipe caused by the higher velocity in the pipe at the higher flowrate. As the flow in a pipe changes, the energy loss due to friction changes. The important point to be emphasized here is that a pump puts head into the system.

A more common situation is illustrated in Figure 11–3. It is clear that water will not flow by itself from reservoir A to B. The pump shown must overcome the friction losses in the pipe as well as the difference in elevation (Z_2) between the water surfaces in the reservoir to deliver water from point 1 to 2. The total head against which the pump is working is the sum of friction head and the static head in feet. A pump may provide head for any purpose. If reservoir B in Figure 11–3 were removed, and a fire nozzle inserted in its place on the end of the pipe, it is easy to visualize the pump supplying elevation head, friction head, and velocity head. In fact, if the pump were placed above the reservoir and properly primed, it could supply the heads enumerated above plus generate a suction head to lift the water up to the pump setting.

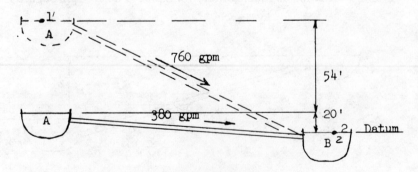

Figure 11–2.

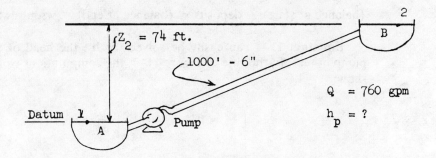

Figure 11–3.

Mathematically, the entire operation depends on adding all the different heads which are required and selecting a pump which can supply this total pumping head. This head and the desired flow constitute the specified head-discharge of the pump. These are sometimes referred to as the rated head and capacity of the pump.

Again, the computation is easily visualized using Bernoulli's equation. Referring again to Figure 11–3, and placing the pump close enough to reservoir A to neglect piping between the reservoir and the pump, and setting up Bernoulli's equation from point 1 to 2, the following may be deduced:

$$\frac{p_1}{w} + \frac{V_1^2}{2g} + Z_1 + h_p = \frac{p_2}{w} + \frac{V_2^2}{2g} + Z_2 + (H_L)_{(1-2)} \quad (11.1)$$

$$\underset{\substack{\text{energy} \\ \text{at 1}}}{\longleftrightarrow} + \underset{\substack{\text{energy} \\ \text{gained}}}{\leftrightarrow} = \underset{\substack{\text{energy} \\ \text{at 2}}}{\longleftrightarrow} + \underset{\substack{\text{energy} \\ \text{lost}}}{\longleftrightarrow}$$

There is a difference in this equation from those used up to this time. If the student recalls, this equation is an energy equation and sums up all energy at point 1 as being equivalent to all energy at point 2 plus the energy loss from 1 to 2.

When energy is put into the system, it must be added to the left side of the equation and when energy is taken out it must be added to the right side of the equation to create a balance. This is similar to a bank balance.

(balance at start) + deposits = (balance at end) + withdrawals

Equation 11-1 can easily be solved for h_p, the head of the pump in feet. In the case of Figure 11-3, the computations would show:

$$
\text{energy at 1}
\begin{cases}
\dfrac{p_1}{w} = 0 \\[2mm]
\dfrac{V_1^2}{2g} = 0 \\[2mm]
Z_1 = 0
\end{cases}
\qquad
\text{energy at 2}
\begin{cases}
\dfrac{p_2}{w} = 0 \\[2mm]
\dfrac{V_2^2}{2g} = 0 \\[2mm]
Z_2 = 74 \text{ ft}
\end{cases}
$$

$$0 + 0 + 0 + h_p = 0 + 0 + 74 + (H_L)_{(1-2)}$$

From Figure 5-2, $(H_L)_{(1-2)} = 74$ ft (neglecting minor losses) therefore:

$$h_p = 74 + 74$$

$$h_p = 148 \text{ ft}$$

In many manufacturers' catalogs, the above equation would merely be stated as "the pumping head is equal to the lift plus the losses." Bernoulli's equation is one way to make certain that no portions of the problem are neglected.

PUMPING MECHANISMS

To supply the energy required to pump fluids, there are two main classifications of pumps, the positive displacement pump and the centrifugal pump.

As the name implies, the positive displacement pump moves the water by positively displacing a volume. This is done using a piston in a cylinder or a flexible diaphragm in a chamber in combination with a system of check valves.

The centrifugal pump relies on centrifugal force, which is de veloped by spinning or rotating an impeller at a high speed, similar to twirling a weight at the end of a rope or a pail of water at the end of a rope. This force throws the water to the outside of the pump with a high velocity. At this point a volute casing, much like the casing on the venturi meter, converts velocity head to pressure head. The emphasis of the discussion will be on the centrifugal pump since it is more commonly encountered in water and sewage practice (Figure 11–4).

In its basic form, the pump consists of an impeller, which constitutes the moving portion of the unit. This, in turn, rotates in an enclosed housing or bowl. Water flows into the mechanism through the side of the housing, into the center of the impeller and, as the impeller rotates, the water is accelerated and thrown with a high velocity to the outside of the casing, or housing. At this point, the velocity is reduced, thus increasing the pressure, and the water is forced out of the pump. As the water is thrown

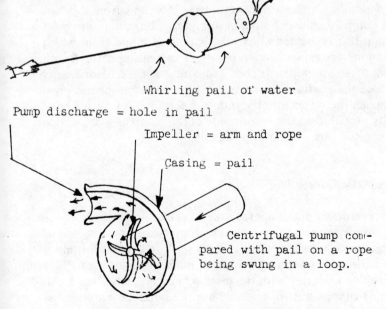

Whirling pail of water

Pump discharge = hole in pail

Impeller = arm and rope

Casing = pail

Centrifugal pump compared with pail on a rope being swung in a loop.

Figure 11–4. Principle of the centrifugal pump.

away from the center of the impeller, a low pressure is produced at the center or eye of the impeller. More water then is drawn into the low pressure area through the suction piping and the flow continues, producing a continuous pumping action. The pressure developed depends on the speed of rotation, design and diameter of the impeller. It is not due to any impact or displacement. The effect of one impeller may not be adequate or sufficient to lift the water as high as necessary, and this will give rise to a need for multiple impellers. This will be discussed in Chapter 12.

The parts of the pump, then, are the impeller, which is the heart of the pump; the casing, bowl, or housing that is around the impeller; and some means of turning the impeller, such as an electric motor and shaft. The shaft has a stuffing box or watertight connection which prevents leakage outside the pump as the impeller rotates.

In the centrifugal pump, there are no valves or reciprocating parts. The impeller and shaft assembly are the only accessories that move. There are no close tolerances or rubbing surfaces within the pump except perhaps some wearing rings which can be replaced when they indicate excessive wear. The flow from the centrifugal pump is a smooth nonpulsating flow, but the pump must be primed if it is located above the source of liquid from which it is pumping (negative suction head). If the casing is filled with air or vapor, the impeller will not pump out such gases sufficiently to produce the partial vacuum needed for the atmospheric pressure to push the water into the pump. Therefore, a positive head is usually provided on the suction side of the pump or some source of priming the pump must be available.

Characteristic Curves

Since a centrifugal pump operates without close tolerances, some slipping of water from the point of high pressure to the region of low pressure is possible. This can be increased by throttling a valve on the pump discharge. In fact, many types of centrifugal pumps can be operated with the discharge completely closed. All centrifugal pumps thus operate with a potential for varying discharge and corresponding head. When a centrifugal pump is operated at a relatively constant speed, which is the most common type of operation, the head against which the pump is working, in feet, and the discharge from the pump, in gallons per minute, can be plotted as shown in Figure 11–5. This curve is known as the head-discharge curve for the pump. On this same graph is usually shown the efficiency of the pump as well as the brake horsepower. The combination of these three curves is known as the characteristic curve for the pump. The efficiency and brake horsepower are on the right vertical axis, the head against which the pump operates is on the left vertical axis and the flow, Q, is on the horizontal axis.

The efficiency of a pump is computed by dividing the work done by the pump, by the work required to operate the pump. Each value of efficiency is plotted against the corresponding discharge. The brake horsepower (the power applied to the pump by the motor or drive mechanism) is also plotted against the appropriate value of discharge.

As will be noted in looking at this set of characteristic curves (Figure 11–5), it is very common to find that as the head changes, a change in discharge will occur. As the head against which the

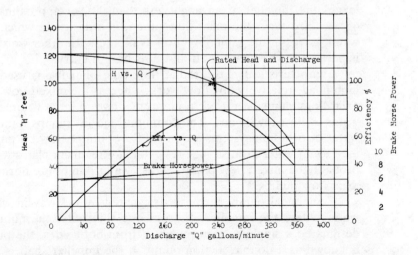

Figure 11–5. Characteristic curve: centrifugal pump at constant speed. As seen in the curve, the maximum efficiency for this pump is 80%, its rated head is 98 ft and its rated discharge is 240 gpm.

pump is working is increased, the discharge decreases; as the head against which the pump is working decreases, the discharge will increase and, of course, as the discharge increases the power requirements will also increase.

There is a maximum point, or peak, on the efficiency curve which shows the point at which the pump is most efficient and this will be the point where the pump should be most commonly operated. As the discharge varies from the optimum discharge-head relationship, the efficiency will be reduced. Operation away from the point of peak efficiency is a more expensive way of operating the pumping unit.

Types of Centrifugal Pumps

There are several types of centrifugal pumps. A pump may be classified by the direction of flow as it leaves the impeller. The orientation most frequently encountered is one in which the flow leaves the impeller at 90° from the direction at which it entered the impeller. This is known as the radial flow pump. The other designs are the so-called mixed flow pump where the discharge leaves the impeller at an angle less than 90°, or approximately 45°; a third type, the axial flow pump, is a design in which the water leaves in the same direction as it enters. In other words, it is parallel to the pump shaft and direction of entry.

Characteristically, the radial flow type of pump is used for high head, relatively low flow work, whereas the mixed flow and axial flow pumps are used to pump medium to large volumes of water against low heads. The common high lift pumps found in water distribution systems are radial flow pumps. Low lift pumps used to lift water from storage reservoirs to treatment plants would probably be mixed flow pumps. Wash water pumps are normally axial flow pumps.

Centrifugal pumps are also designated by the way in which water enters the impeller. If from one side only, the pump is designated a single suction pump. If from both sides, the pump is known as a double suction pump. If the impeller shaft is in a vertical position, the pump is called a vertical centrifugal pump. If it is in a horizontal position, it is called a horizontal centrifugal pump.

It should be noted in the case of the centrifugal pump, by referring to the characteristic curve (Figure 11–5), that at zero discharge there is a certain head or pressure developed in the pump. This is commonly referred to as the "shutoff head" and is usually the maximum pressure that the pump can develop. Centrifugal pumps of the radial or mixed flow variety are designed such that usually no harm is done if the discharge valve on the pump is closed when the pump is turned on. As a matter of fact, this is not an uncommon way for these pumps to be put into operation since the power requirements are lowest at that time. As the valve is opened, the flow begins and the pump comes up to normal operating conditions. This is quite different from the operation of a positive displacement pump in this respect. Axial flow pumps are not used in this fashion either, since the head-discharge curve is so steep and the power requirements so high at shutoff head that there is danger in stalling the prime mover.

Centrifugal pumps operate in a manner described by their characteristic curves and do not stray from them. It is not uncommon to want to know what a pump will do when placed in an existing pipe system. Since the discharge of the pump depends on the head against which it is working while the head loss in the pipe system depends on the discharge through it, it becomes necessary to make assumptions in a trial and error type of solution.

Example 11–1

Calculate the discharge from a pump whose characteristic curves are shown in Figure 11–5. The static lift is 75 ft. The pump discharges into a main already carrying 300 gpm. The main is 6 in in diameter and 1000 ft long. Assume C = 100.

First Trial

Head loss in the main for flow of 300 gpm, H_L:

(Figure 5–2) H_L = 13.0 ft

Static lift   = 75.0 ft

Total pump head (h_p) = 88.0 ft

At 88 ft head, pump will deliver 280 gpm (from Figure 11–5)

Second Trial

Assume 220 gpm from pump

$$\text{Total flow 520 gpm } (H_L) = 35.0 \text{ ft}$$

$$\text{Static lift} \quad\quad\quad = \underline{75.0 \text{ ft}}$$

$$\text{Total head} \quad\quad\quad = 110.0 \text{ ft}$$

$$Q \text{ from Figure 11–5} = 170 \text{ gpm}$$

Third Trial

Assume 200 gpm from pump

$$\text{Total flow 500 gpm } (H_L) = 34.0 \text{ ft}$$

$$\text{Static lift} \quad\quad\quad = \underline{75.0 \text{ ft}}$$

$$\text{Total head} \quad\quad\quad = 109.0 \text{ ft}$$

$$Q \text{ from Figure 11–5} = 180 \text{ gpm}$$

Fourth Trial

Assume $Q = 185$ from pump

$$\text{Total flow 485 gpm } (H_L) = 33.0 \text{ ft}$$

$$\text{Static lift} \quad\quad\quad = \underline{75.0 \text{ ft}}$$

$$\text{Total head} \quad\quad\quad = 108.0 \text{ ft}$$

$$Q \text{ from Figure 11–5} = 185 \text{ gpm}$$

This is the discharge assumed. Therefore, the new pump would operate at a head of 108 ft and discharge of 185 gpm. This corresponds to 72% efficiency, which is about 8% off the peak value.

POSITIVE DISPLACEMENT PUMPS

The positive displacement pump, as indicated earlier in the chapter, actually pumps water by positively moving a given volume in a closed container, such as a piston in a cylinder. The operation of a positive displacement pump results in a pulsating flow in most cases. Its large number of moving parts due to the required inlet-outlet valves, timing cams, reciprocating equipment, etc., gives rise to the greater complexity of the reciprocating pump in general. The exceptions might be the rotary positive displacement pump which consists of closely fitting cams or gears which run together inside a casing and literally squeeze the water through the pump, or the diaphragm pump which has a very simple rubber diaphragm which is raised and lowered in a chamber. The characteristics of a positive displacement pump are such that at shutoff or zero discharge tremendous pressures can be developed. As a result, failure of either the pump, prime mover, or pipe may occur if the discharge valve is closed when the pump is started or in operation unless a bypass or pressure release valve is available.

ADDITIONAL EXAMPLES USING CHARACTERISTIC CURVES

Example 11–2

If a pump with characteristic curves as in Figure 11–5 is operating against a head of 100 ft, what will be its discharge and efficiency? What horsepower will be needed to operate it?

Answer:

At 100 ft of head, this pump will deliver 235 gpm with an efficiency of 79% and a brake horsepower requirement of 7.3 hp.

Example 11–3

If the head in the example above changes to 110 ft, what discharge, efficiency, and horsepower will result?

146   *Hydraulics for Operators*

Answer:

At 110 ft of head, the flow will be 165 gpm, efficiency 67%, and power required 6.9 hp.

Example 11-4

If the head changes to 60 ft, what will be the discharge, efficiency, and power required?

Answer:

Discharge = 350 gpm, efficiency = 45%, 11 hp.

Example 11-5

What is the shutoff head of this pump?

Answer:

120 ft

Example 11-6

If the efficiency of this pump is to be held at no less than 70%, what flows can be pumped? Against what head?

Answer:

Flows ranging from 180 gpm to 290 gpm and with heads ranging from 108 ft to 85 ft.

PROBLEMS

11-1.   It is desired to pump a flow of 200 gpm with the pump whose characteristics are shown in Figure 11-5. What head will this flow be delivered against and what efficiency and brake horsepower (bhp) can be expected?

Answer: Head = 105 ft
Efficiency = 75%
bhp = 7

11-2. A pipeline 500 ft long connects two reservoirs. A pump drains water from the first reservoir and discharges into the pipeline. The water surface of the second reservoir is 40 ft above that in the first. A pump with characteristic curves as shown in Figure 11-5 is to be used and it is desired that 300 gpm be pumped. What should be the diameter of the pipe and what will be the head against which the pump will work? Neglect minor losses. C = 100.

Answer: Head = 84 ft
Diameter = 4 in

11-3. A pump whose characteristics are similar to Figure 11-5 is to pump into a 10-in main 2000 ft long with an existing flow of 1000 gpm. The static lift is 60 ft. Compute the discharge from the pump. C = 100. Neglect minor losses.

Answer: Q = 275 gpm

11-4. From a reservoir whose surface is at elevation 750, water is pumped through 4000 ft of 12-in pipe across a valley to a second reservoir whose level is at elevation 800. If, during pumping, the pressure is 80 psi at a point on the pipe, midway of its length, and at elevation 650, compute the rate of discharge and the power exerted by the pumps. C = 100. Neglect minor losses.

Answer: Q = 2200 gpm
Total head = 121 ft

11–5.   A pump is drawing water from a tank through a 4-in
smooth suction pipe. The rate of pumping is 500 gpm.
Compute the pressure in psi at point B on the suction side
of the pump. C = 100.

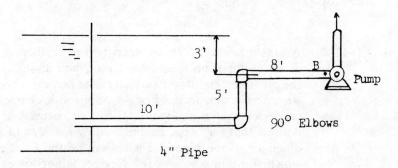

Answer: Pressure at B = −4.4 psi.

12

Pumping

In this chapter, discussion will be concerned only with the use of centrifugal pumps. This will include a discussion of pumps in series and in parallel, and the concept of suction lift and the computation of horsepower requirements. As has been discussed in Chapter 2, atmospheric pressure amounts to about 14.7 psi at sea level. This means that the pressure of the atmosphere pushing down on the surface of a liquid will push that liquid, under theoretical conditions, to a height of 34 ft. Thus, again theoretically, a pump can be located as high as 34 ft above the water being pumped and, after being primed, lift water that distance. In actual practice, due to atmospheric conditions, losses due to friction, leaks in the pipe, vaporization of water, and other reasons, the practical suction lift with a properly primed centrifugal pump is no more than 15 ft. It is best, as indicated in the preceding chapter, to provide water to the centrifugal pump at the suction side under a positive pressure so that the pump can be self-priming. If a suction lift is to be developed, the centrifugal pump must have a way of being primed. This means that the casing of the pump must be filled with water by some means before the pump is started.

HORSEPOWER

For computing pumping power requirements, the equation relating horsepower to flow, head, and pump efficiency can be used in the following form:

149

$$hp = \frac{gpm \times pumping\ head}{3960 \times E}$$

where gpm = amount of water being pumped (previously
called Q), gal/min

pumping head = summation of all heads against which the
pump is working (obtained from Bernoulli's
equation, called h_p in Chapter 11), ft

E = efficiency of the pump (obtained from
the efficiency curve for that pump)

The horsepower thus determined will be the horsepower needed
to drive the pump under these conditions and can be used to select
the motor needed.

It must be cautioned that a value for horsepower can be obtained for any of the head-discharge relationships which lie along
the head-discharge curve of the pump (see Figure 11–5). In selecting a motor, obviously one must be chosen which will satisfy
the worst case which the pump may be called upon to provide in
actual service. The rated head-discharge value may not produce
the critical condition. This can best be defined by close study of
the pump's characteristics.

Example 12–1

A pump discharging 500 gpm from one reservoir to another is
operating against a head of 120 ft. At this point, the efficiency
from the characteristic curves is 68%. What is the horsepower
required to operate this pump?

$$hp = \frac{gpm \times head}{3960 \times E}$$

$$hp = \frac{500 \times 120}{3960 \times 0.68} = 22\ hp$$

PUMP AND PIPE COMBINATIONS

It is often necessary in practice to predict flow for a system with a known pump and an existing piping system. This involves some trial and error plotting on the pump "head vs flow" characteristic curve, and requires consideration of many factors previously discussed in this text. Use of an example will most clearly illustrate the approach to such a problem.

Example 12–2

A pipe system with conditions as outlined below is already installed in a water treatment plant. If a 4-in centrifugal pump, with characteristic curve as shown in Figure 12–1, is installed, how much water will be pumped? The system has the following pipe arrangement.

Suction Side. 30 ft of 8-in, 15-yr-old cast iron pipe; one 8-in to 4-in reducer; one 8-in foot valve; and one 90°, 8-in-long sweep elbow.

Discharge Side. 2000 ft of 6-in, 15-yr-old cast iron pipe; one 6-in gate valve; one 6-in to 4-in reducer; and five 90°, 6-in-long sweep elbows.

Static Head. Suction side with 5 ft of lift, discharge side with 20 ft of head.

Solution

Assume about three different flowrates at random (say, 500, 750, and 1000 gpm). Next, calculate the total head for each condition of flow in the existing pipe system and plot on the characteristic curve (Figure 12–1). An example of calculations for the 750 gpm flow follows:

Suction.	
Feet of 8-in pipe	30 ft
Length of 8-in pipe equivalent to 1–90°, 8-in-long sweep elbow	14 ft

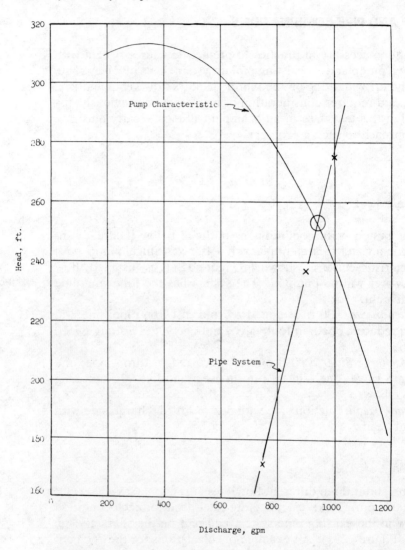

Figure 12–1. Characteristic curve for 4-in pump.

Length of 8-in pipe equivalent to
  1-foot valve   10 ft
Length of 8-in pipe equivalent to

1 reducer (d/D = 1/2)   4 ft

Suction equivalent 8-in piping; Total =   58 ft

Discharge.
Feet of 6-in pipe   2000 ft
Length of 6-in pipe equivalent to
5–90°, 6-in-long sweep elbows (5 × 11)   55 ft
Length of 6-in pipe equivalent to
1 reducer (d/D = 2/3)   3 ft
Length of 6-in pipe equivalent to one
6-in gate valve (wide open)   3.5 ft

Discharge equivalent 6-in piping; Total = 2061.5 ft

Then, from Figure 5–2, loss of head due to friction can be calculated for the equivalent lengths of 8-in and 6-in pipe:

8-in suction friction head	1.0 ft
6-in discharge friction head	146.0 ft
Suction lift, static	5.0 ft
Discharge lift, static	20.0 ft
Total dynamic head	172.0 ft

For flows of 500 and 1000 gpm, the total head may be calculated in similar fashion to be 85.2 and 275 ft, respectively. Since 500 gpm gives a lower head than can be read on Figure 12–1, it is well to calculate a point intermediate between 750 and 1000 gpm. If, for example, 900 gpm is chosen, one can calculate the total dynamic head to be 237 ft.

Then, if all three head values are plotted as shown on the pump characteristic curve, it may be observed that with this pump and this system, the flow will be about 940 gpm. This value is defined by the point where the pipe system curve crosses the pump characteristic curve. Lesser flows may be obtained by adding head loss to the piping system. This may be done easily by throttling a valve. No greater flow than 940 gpm can be obtained through this piping system.

An additional example serves to further illustrate approaches to an overall solution to pump-pipe combinations.

Example 12–3

Given the pump as shown:

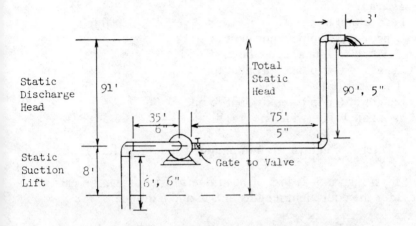

The suction pipe is 6-in in diameter and the discharge pipe is 5-in in diameter. The pump is pumping 500 gpm with an efficiency of 70%.

What is the total head against which this pump is working, and how much power is needed to drive it?

Writing a Bernoulli equation between a point on the water surface at the pump suction and a point at the discharge of the 6-inch pipe gives:

$$H_p = \frac{V_2^2}{2g} + Z_2 + H_{L_{(1-2)}}$$

(All other terms in this Bernoulli equation are zero.) Therefore, the pump must supply energy to meet the velocity head ($V_2^2/2g$), the static head (Z_2), and the friction head ($H_{L_{(1-2)}}$).

Static Head

Static suction lift	8 ft
Static discharge head	91 ft
Total static head	99 ft

Pressure Head

Since both suction and discharge water surfaces are open to the atmosphere, there is no pressure head.

Head required due to pipe friction:

Six-inch Pipe. First, fittings are converted to equivalent length of straight pipe.

6-in Borda entrance	16 ft
6-in, 90° long radius elbow	11 ft
Straight 6-in pipe	6 ft
Straight 6-in pipe	35 ft
Total equivalent length of 6-in pipe	68 ft

Five-inch Pipe.

Fittings

5-in gate valve, completely open	3 ft
Two 5-in, 90° long radius elbows (2 × 9)	18 ft
165 ft straight 5-in pipe	165 ft
Total equivalent length of 5-in pipe	186 ft

Friction Head.

Q = 500 gpm (use nomograph in Chapter 5)
6-in pipe, 68 ft long, H_L =
$\quad$ 0.034 ft/ft = 0.034 × 68 $\qquad$ 2.32 ft
5-in pipe, 186 ft long, H_L =
$\quad$ 0.08 ft/ft = 0.08 × 186 $\qquad$ <u>14.9 ft</u>
Total Friction Head = $\qquad$ 17.2 ft

Velocity Head.

$$A = \frac{\pi(5/12^2)}{4} = 0.136 \ ft^2$$

$$Q_{cfs} = \frac{Q_{(gpm)}}{60 \times 7.48} = \frac{500}{60 \times 7.48} = 1.11 \ cfs$$

$$V = \frac{Q}{A} = \frac{1.11}{0.136} = 8.16 \ ft/sec$$

$$\text{Velocity head} = \frac{V^2}{2g} = \frac{8.16 \times 8.16}{64.4} = 1.04 \text{ ft}$$

Total head against which the pump is working:

Static head	99.0 ft
Pressure head	0
Friction head	17.2 ft
Velocity head	1.04 ft
	117.24 ft

Since velocity head is usually a small amount of head loss with respect to the total head, it generally is neglected.

Power Required (equation from Chapter 11)

$$\text{Horsepower} = \frac{\text{gpm} \times \text{head}}{3960 \times \text{E}} = \frac{500 \times 117.2}{3960 \times 0.7} = 21.2 \text{ horsepower}$$

CENTRIFUGAL PUMPS IN SERIES

If the head against which a pump is to work is greater than the head that can be developed by one impeller, then it is common practice to put more than one impeller together in series. When this is done, the discharge of one impeller is directed into the eye of the next impeller and on into another, depending upon the number required so that the flow is constant through each impeller, but the head supplied by each one is added to that of the previous impeller. As a result, this multistage pump will deliver water to as high a head as is necessary. The system, head vs discharge, curve of more than one impeller in series is best shown by a diagram with the discussion of how this system curve is produced (Figure 12–2). It should be noted that impellers in series implies more than one impeller in a single casing. Pumps in series implies two or more pumps connected in series. The same theory applies in both cases. The arrangement of impellers in series is best illustrated in practice by the deep well turbine pump which may consist of several stages or impellers on one shaft driven by one motor.

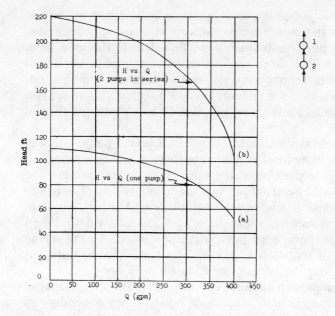

Figure 12–2. H vs Q curves for (a) one pump and (b) two pumps with characteristics like (a) in series.

Example 12–4

If the head against which two pumps in series, for which characteristic curves are shown in Figure 12–2, are working against a head of 200 ft, what will be the discharge from the two pumps? Refer to Figure 12–2.

Answer: 190 gpm

PUMPS IN PARALLEL

When it is necessary to pump greater amounts of water than the capacity of one pump alone, it is common practice to put two or more pumps in parallel, in which case all pumps take water from the same source and deliver it to a common header or discharge line. When pumps are connected in this way, it is approximately correct to assume that the head against which each of the pumps is working is the same, but as more pumps are added the discharge pressure normally increases. To calculate the total flow, the individual flows from each of the several pumps are added together to obtain the total flow from the parallel pumping setup. The development of the system, head vs discharge, curve for two pumps in parallel again, is best illustrated by an example (Figure 12–3).

If two pumps with identical characteristics are put together in series, the resulting head vs discharge curve would be determined as follows: Since the flow is the same through both pumps, the head is supplied by each of the various flows. At zero (Figure 12–2), pumps 1 and 2 have a head (shut-off) of 110 ft, the system (both in series) then has a shut-off head of 220 (110 + 110). When the flow is 100 gpm, each pump develops 106 ft of head so the system head at that flow is 212 ft. This type of computation is used to complete the system curve shown in Figure 12–2.

If two pumps with identical characteristics are put together in parallel, discharging into a common header, they generally can be considered to be operating against a common head. The H vs Q curve for the two pumps in parallel (system curve) is computed by assuming various heads and adding the flows produced by each pump at that head to get the total discharge of two pumps acting together. At 110 ft of head (shut-off head) neither pump is discharging anything (Figure 12–3). At 100 ft of head, each pump is delivering 290 gpm for a total of 580 gpm (290 + 290). At 80 ft of head, each pump discharges 392 gpm, a total of 784 gpm, and so on. For the computation of the system H vs Q curve, several such points are taken and the resulting curve drawn.

Example 12–5

Two pumps connected in parallel as shown in Figure 12–3 are pumping against a head of 90 ft. What is the total flow into the system? Refer to Figure 12–3.

Answer: Each pump discharges 350 gpm, a total of 700 gpm.

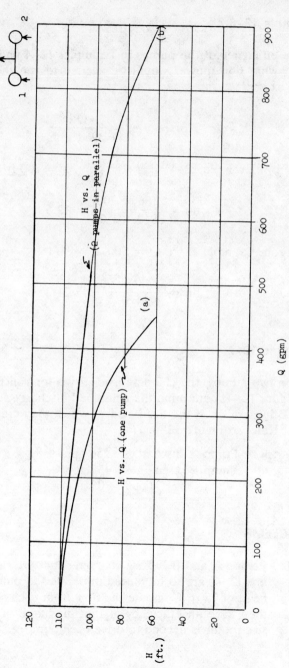

Figure 12–3. H vs Q curves for (a) one pump and (b) two pumps with characteristics like (a) connected in parallel.

Example 12–6

If the efficiency of the pumps in Examples 12–4 and 12–5 were 75%, what horsepower would be needed to run the system in each example?

$$\text{Horsepower} = \frac{Q\,h}{3960 \times E}$$

From Example 12–4, Q = 186 gpm, h = 200 ft

$$\text{hp} = \frac{186 \times 200}{3960 \times 0.75} = 12.5$$

From Example 12–5, Q = 700, h = 90 ft

$$\text{hp} = \frac{700 \times 90}{3960 \times 0.75} = 21.2$$

Example 12–7

Given two pumps, the characteristic curves for which are shown in Figure 12–4, determine the combined discharge for pumps A and B in parallel, if the head against which they are pumping is 70 ft. Refer to Figure 12–4.

Answer: Pump A discharges   250 gpm
Pump B discharges   455 gpm
Total   705 gpm

PROBLEMS

12–1. Pumps A and B, whose characteristics are shown in Figure 12–4, are to be placed in parallel to pump against a head of 75 ft. Compute the flow from each pump, and if they have an efficiency at that discharge of 70%, what size motor is needed to drive each pump?

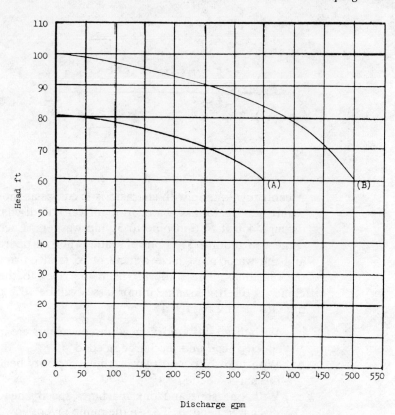

Figure 12-4.

Answer: Pump A = 175 gpm, Horsepower$_A$ = 5
Pump B = 425 gpm, Horsepower$_B$ = 12

12-2.   Pumps A and B of Figure 12-4 are to be placed in series. If the flow desired is 150 gpm, what head will this flow be delivered against, and what head is contributed by each pump?

Answer: Pump A = 150 gpm at   76 ft
Pump B = 150 gpm at   95 ft
Total head   171 ft

12–3.

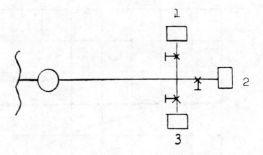

A centrifugal pump with characteristic curves as shown in Figure 12–1 is to take water from a river and discharge it through 450 ft of 6-in pipe to 3 pulp washers. Each machine will require 350 gpm of water when in operation and the water must have a head of 65 ft at entrance to the machines. The static lift from the river to the machines is 100 ft. (Assume minor losses will be 10% of the pipe loss and C = 100.)

a. With three shifts working all three washers are in use, specify head and discharge needed.
b. With two shifts and two machines, specify head and discharge needed.
c. With one shift and one machine, specify head and discharge needed. To use the pump given, how much head must be lost in throttling valve?

Note: The student should realize that this pump would be a poor choice for such service. If one pump must be used, it should have a flat head-discharge curve.

Answer: a. 239 ft, 1000 gpm (from curve)
           b. 195 ft (throttled), 700 gpm
           c. 173 ft, 350 gpm, 137 ft of head lost

Index